AF881062

SOTHIS PRESS

Sothis Press aims to completely reimagine the public domain book with an emphasis on universal access to captivatingly designed print editions. We believe that technology will enable every single book ever published to be perpetually accessible in beautiful print form. Sothis is beauty in ubiquity.

www.sothis.press
1 Waterhouse Square, London, ECIN 2ST

Watchmakers' and Jewelers' Practical Receipt Book

A Workshop Companion; Comprising Full and Practical Formulae and Directions for Soldiers and Soldering, Cleaning, Pickling, Polishing, Coloring, Bronzing, Staining, Cementing, Etching, Lacquering, Varnishing, and General Finishing of Metals

C. E. Walker

SOTHIS PRESS

ISBN 978-1-0368-0723-8

Watchmakers' and Jewelers' Practical Receipt Book
is in the public domain.

This edition is published by
Sothis Press © 2023. The company
is registered in England and Wales.
The Company Number is 14900389.

WATCHMAKERS' AND JEWELERS'

PRACTICAL RECEIPT BOOK

A WORKSHOP COMPANION.

COMPRISING FULL AND PRACTICAL FORMULAE AND DIRECTIONS FOR SOLDERS AND SOLDERING, CLEANING, PICKLING, POLISHING, COLORING, BRONZING, STAINING, CEMENTING, ETCHING, LACQUERING, VARNISHING, AND GENERAL FINISHING OF METALS, AS APPLIED TO THE WATCH AND JEWELRY TRADE. TOGETHER WITH ALL THE IMPORTANT ALLOYS USED BY THE TRADE AND MANY MISCELLANEOUS RECEIPTS.

COMPILED FROM PRIVATE FORMULAE.

CHICAGO:
GEO. K. HAZLITT & CO., PUBLISHERS.
1892.

WATCHMAKERS' AND JEWELERS'

PRACTICAL RECEIPT BOOK.

CEMENTS.

Amber Cement.

1. Moisten the broken edges of the amber with a solution of potash and press them together and keep in this position until dry.

2. Heat slightly the broken surfaces, apply a small portion of shellac to the warmed surface and press closely together. After all is dry remove the superfluous cement by means of a sharp knife and polish with a flannel rag and oil.

3. Heat the broken surfaces, apply boiled linseed oil and press firmly together until dry.

4. Mastic melted in boiled linseed oil and applied to the broken parts is said to make a perfect joint.

Jeweler's Armenian Cement.

1. Melt one part of thick isinglass glue, mix with one part of thick mastic varnish and keep in a well corked bottle. To use, heat in hot water.

2. Soak isinglass in water and dissolve in 2 oz. of spirit to form a thick paste; dissolve 10 gr. of pale gum ammoniac in this by rubbing well together. Add to this

six large tears of gum mastic dissolved in the smallest possible quantity of alcohol.

3. Dissolve six pieces of gum mastic the size of a pea in as much spirits of wine as will suffice to render it liquid. Dissolve two oz. of isinglass in water, pour off the superfluous liquor and dissolve in rum, adding 10 gr. of gum ammoniac, which must be thoroughly incorporated by rubbing in with the liquid until dissolved. Mix this with the gum mastic, using heat. Put up in a closely stoppered bottle and when about to use heat the bottle in hot water. This cement is said to be very effective in uniting all substances, even glass to polished steel.

4. Keller's Armenian cement. Soak for twenty-four hours ½ oz. of isinglass in 4 oz. of water; evaporated in a water bath to 2 oz; add 2 oz. alcohol and strain through a linen cloth. Form a solution of ¼ oz. best mastic and 2 oz. alcohol and mix while warm with the above. Add to this 1 drm. gum ammoniac and mix until thoroughly incorporated. Avoid the loss of the spirit by evaporation as much as possible.

5. Ure's Armenian cement. Water 6 oz., isinglass 1 oz; boil to 3 oz., and add 1½ oz.; rectified alcohol, boil for a minute or two, strain and add while hot, first a milky emulsion of ammoniac, ½ oz., then 5 drm. tincture of mastic.

Acid Proof Cement.

1. Quicklime and linseed oil mixed to a stiff paste forms a hard cement which resists both acid and heat.

2. India rubber melted by gentle heat is mixed with from 6 to 8 per cent. by weight, of tallow, stirring well the while; dry slacked lime is now added until the mass assumes the consistency of a thick paste; now add 20 per cent. of red lead which will make it harden and dry.

3. Mix a concentrated solution of silicate of soda with powdered glass to form a paste.

4. Melt 25 grm. of old gutta percha, previously reduced to shreds, and mix with 75 grms. of powdered pumice and then add 150 grms. of Burgundy pitch. This cement will be found useful for lining troughs used in galvanoplastic manipulations and will resist sulphate of copper baths but not cyanides.

5. Melt one part India rubber with two parts linseed oil; add sufficient white bolus for consistency. Neither muriatic nor nitric acid attack it; it softens a little in heat, and its surface does not dry easily; which is produced by adding one-fifth part of litharge.

6. Sulphur 100 parts, resin 2 parts, tallow 2 parts Melt and add sifted ground glass until brought to the proper consistency.

Alabaster Cement.

1. Finely powdered plaster of Paris made into a paste with water. This cement may be used to join and to fit together pieces of alabaster or marble, or to mend broken plaster figures.

2. Melt rosin, or equal parts of yellow rosin and beeswax, then stir in half as much finely powdered plaster of Paris. This cement is used to join alabaster, marble, and other similar substances that will bear heating.

3. Melt alum and dip the fractured faces into it; then put them together as quickly as possible. Remove the exuding mass with a knife.

4. To ½ pint skimmed milk add ½ pint vinegar and mix the curd with the white of five eggs well beaten, and sufficient powdered quick-lime sifted in, stirring constantly, to form a thick paste.

5. Prepare a thin paste by boiling rice to a pulp or using rice flour and hot water and thicken this paste with finely powdered quick-lime. Slightly heat the fractured surfaces, apply and place in a warm spot until dry.

Cement for Leather.

1. Take equal parts of isinglass and glue and add enough water to cover. Let this soak for ten hours and then bring to a boil and add pure tannin until the mixture becomes stringy or like the white of eggs. Rough up the surfaces of the leather to be united and apply the mixture hot to them. Place the joint under pressure of heavy weights for a few hours and it will be found that a joint has been made as strong as the leather itself. This will be found to be an excellent cement for joining flat lathe belts, draw skins, etc.

2. Shred or cut into small pieces pure India rubber. Select a wide mouthed bottle with glass stopper and fill about one-tenth full of the shreds and fill the bottle with pure benzine, which is free from all oil. Allow the mixture to stand until the benzine has thoroughly cut or dissolved the rubber. If too thin, add rubber and if too thick, benzine. This cement will be found excellent for fastening small belts or joining other pieces of leather or rubber.

Cement for Bisque Figures.

Carefully wash and dry several oyster shells, burn well, slack in the air and reduce to a fine powder in a mortar. Pass through a fine sieve and mix with the white of egg. Clean thoroughly the fractured surfaces, heat slightly and apply the cement keeping the parts well pressed together for a few minutes until dry.

Label Cement.

1. Gum Arabic 2 oz., gum tragacanth ½ oz., water ½ pint. Stir until thoroughly dissolved, then strain and add 7 grains thymol, 2 oz. glycerine and water to make one pint.

2. Half fill a wide-mouthed glass stoppered bottle with acetic acid and add isinglass until it forms a thick paste, which it will in a few hours. This cement is useful for attaching labels to tin or glass, which it does very effectually.

3. Alcohol 95 per cent. 10 oz., ordinary cooking gelatine 1 oz., glycerine 1 oz. The gelatine should be previously soaked in cold water for an hour and the superfluous moisture squeezed out. Preserve in wide-mouthed glass-stoppered bottle. This cement is also valuable for mounting photographs, and will not discolor or turn yellow.

4. One or two drops of glycerine added to a small bottle of mucilage is said to prevent the mucilage drying too hard, and labels on glass can be kept firmly in place with it.

Coral Cement.

Warm coral very carefully, and with a pencil brush cover the crack with watch oil; when cool the seam will not show,

Cement for Acid Bottles.

Melt 2 parts of tallow over a sand or water bath, and gradually add 30 parts of pure rubber shredded up. After all is thoroughly melted add two parts of slacked lime.

Glass and Metal Cement.

Brass letters, and other articles of a like nature, may be securely fastened on glass windows with the following:

1. Litharge 2 parts, white lead 1 part, boiled linseed oil 3 parts, gum copal 1 part. Mixed just before using. This forms a quickly-drying and secure cement.

2. To 16 parts of copal varnish add 2 parts of turpentine, 5 parts of drying oil, 5 parts of liquid glue and 10 parts stucco.

3. Knead together rosin soap with half the quantity of zinc white.

Gutta Percha Cement.

Resin 2 parts, gutta percha 4 parts and tar 2 parts. Heat in water bath and apply warm. This cement is said to make a perfect joint for gutta percha.

Cutlery Cements.

These cements will be found very effective for fastening blades of knives into ivory or pearl handles. Fill the cavity in the handle with the cement, heat the tang of the blade, crowd it in and remove the superfluous cement.

1. Melt together 5 parts pitch, 1 part hard tallow and 1 part wood ashes.

2. Melt 4 parts resin and 1 part beeswax and stir in 1 part of sifted brick dust.

3. Melt together 4 lbs. resin and 1 lb. beeswax and stir in 1 lb. whiting.

Cements for Ivory.

1. Mix together finely powdered quicklime and white of egg. Apply sparingly, clamp firmly and lay aside for 24 hours.

2. Dissolve in 30 parts of water 2 parts of white glue and 1 part isinglass. Strain and evaporate to 6 parts. Add a small piece of gum mastic previously dissolved in alcohol and 1 part of zinc white.

Cements for China, Porcelain and Glass.

1. In 64 parts of chloroform dissolve 1 part of India rubber and then add from 14 to 20 parts of gum mastic in powder. Allow the whole to dissolve for two days, shaking occasionally Preserve in air-tight bottles.

2. Melt together 4 parts of Burgundy pitch, 6 parts sulphur, 2 parts elemi, 1 part shellac and 6 parts of finely powdered kaolin. The fractured surfaces should be thoroughly cleansed and heated before applying cement.

Enamel Dial Cements.

1. Gum copal 50 parts, damar 50 parts, Venice turpentine 55 parts, zinc white 30 parts, ultramarine 1 to 2 parts. Apply hot and after allowing to cool remove the superfluous cement with a sharp knife and polish with a burnisher and beeswax. This will be found an excellent cement for repairing chipped enamelled dials, either for clocks or watches.

2. Scrape some pure white wax, mix it with equal parts of zinc white, melt the mixture over a spirit lamp, and let it cool. For use, warm the dial plate slightly and press the cold cement into the defective place. The cement adheres very firmly and by scraping with a sharp knife acquires a white and lustrous surface. In case the

cement should be too hard add some wax, and if too soft some zinc white. Cleanliness in the manipulation and moderate heating in mixing are the principal points and contribute essentially to the snow white color of the cement.

Cement for Emery Wheels and Sticks.

Melt together equal parts of white resin and shellac. When thoroughly melted add an equal amout of carbolic acid in crystals.

Cements for Celluloid.

1. In 2 parts spirits of camphor, dissolve 2 parts shellac and add to this 7 parts strong alcohol. Apply warm, but do not bring the celluloid with fire as it is very inflamable.

2. Dissolve finely scraped celluloid in 90 per cent. spirits of wine. Slightly warm it in water bath before applying.

Cement for Horn and Bone.

Dissolve 5 parts of mastic in 2 parts of turpentine and and 6 parts linseed oil. This cement will be found useful for mending horn and bone jewelry, knife handles, combs, etc.

Cement for Jet.

Broken crosses, brooches, ear rings, etc., of jet may be repaired with shellac which has previously been smoked or mixed with alcohol and lamp black to give it a black color like the article. The cement should be applied sparingly and the edges warmed before applying.

Cement for Meerschaum.

A solution of water glass, or silicate of soda and casein is prepared to which is added, by stiring into it, a little calcined magnesia. This cement must be used at once. Casein is prepared by allowing well skimmed milk to stand until it curdles, when the casein is filtered out and washed on a paper filter. To simplify the above, fresh cheese may be boiled in water and mixed with slacked lime and ashes, using 10 parts cheese, 20 parts water, 2 parts lime and 2 parts wood ashes. The broken parts should be bound together after cementing and set aside for 24 hours.

Cement for Hard Rubber.

This cement will be found useful for mending hard rubber jewelry, combs, etc. Dissolve bleached gutta percha in carbon bisulphide. Apply and when dry brush over with carbon bisulphide in which sulphur has been dissolved.

Opticians' Cements.

1. Resin 15 oz., and beeswax 1 oz., melted together. Take 4 oz. of whiting in an iron spoon and bring it to a red heat. Allow it to cool partially and while still warm add it to the above, stirring in well.

2. Fill a wide mouthed bottle ⅓ full of shellac and cover it with rectified spirit of wood naphtha. Allow it to stand until it assumes a jelly. This cement and number 1 will be found valuable for holding glasses while grinding and polishing.

Engraver's Cement.

Resin, 1 part; brick dust, 1 part; mix with heat.

Fireproof Cement.

A very tenacious and fire-proof cement for metals is said to be made by mixing pulverized asbestos with waterglass, to be had in any drug store; it is said to be steam tight, and resist any temperature.

Metal Cement.

Take Plaster of Paris, and mix it to proper thickness by using water containing about one-fourth of gum Arabic. This cement is excellent for metal exposed to contact with alcohol.

Strong Cement.

Mix some finely-powdered rice with cold water, so as to form a soft paste. Add boiling water, and finally boil the mixture in a pan for one or two minutes. A strong cement is thus obtained of a white color, which can be used for many purposes.

Gold and Silver Colored Cement.

A cement for filling hollow gold and silver articles consists of 60 parts shellac, 10 parts Venetian turpentine, and 3 parts gold bronze, or silver bronze, as the case may be. The shellac is melted first, the turpentine is then added, and finally, with constant stirring, the gold or silver bronze.

Tortoise Shell Cement.

1. Shellac, 45 parts; mastic, 15 parts; spirits of wine, 90 per cent., 175 parts; and turpentine, 3 parts. This cement will join broken tortoise shell jewelry, knife handles, combs, cigar holders, etc., in a very neat manner and the joint will be so strong that it will be easier to break the material than the joint.

ENAMELS.

Enamel Fluxes for Colored Enamels.

The ingredients are pounded to a fine powder in a stone mortar and then placed in a heated crucible.

To prepare the fluxes, a suitable furnace is used, which must be entirely free from rust and lined up to the cover with fire bricks set in clay so that only the opening for the door remains free. Through a hole in the center of the cover, which is also provided with a cover, the ingredients in the crucible are stored with an iron rod. To secure the crucible, a piece of brick is laid upon the grate. The firing is done either with charcoal alone, or with charcoal mixed with coke. The following are the most important fluxes:

1. Fuse 8 parts of minium, 1½ parts of borax, 2 of ground flint and 6 of flint glass.

2. Fuse 10 parts of flint glass, 1 of white arsenic and 1 of saltpetre.

3. Fuse 1 part of minium and 3 of flint glass.

4. Fuse 9½ parts of minium, 5½ of borax and 8 of flint glass.

5. Fuse 6 parts of flint glass, 7 of the flux prepared according to No. 2, and 8 of minium.

6. Fuse 6 parts of the flux prepared according to No. 4, with 1 of colcothar.

7. Fuse 6 parts of minium, 4 of borax and 2 of powdered flint.

The fluxes prepared as above are cooled off in water, then dried and finally powdered in a stone mortar.

Brown Enamel.

Fuse together 8½ parts of minium, 4 of pulverized flint, and 2½ of pyrolusite, triturate, 1½ parts of this mixture with 1 part of flux 4, and 1½ parts of iron filings.

Green Enamel.

Prepare green frit by fusing together 1½ parts of green pot metal glass, 3 of pulverized flint, 7½ of borax, 1¼ of green oxide of copper, 7½ of minium and 3 of flux 1. Pound the mixture to fine powder in a stone mortar. Triturate with water 5 parts of this frit, ½ of flux 2 and 2½ of flux 6.

Light Red Enamel.

Triturate with water 1½ parts of white lead, 1 part of red sulphate of iron and 3 of flux 1.

Dark Red Enamel.

Triturate with water 1 part of brown sulphate of iron, and 2¼ of flux 7.

Opaque White Enamel.

Calcine in a crucible 1 part of buck's horn shavings until they are entirely white, and rub them to a paste with 1 part of flux 1. Triturate with water 1 part of Venetian white enamel and 1 of flux No. 8, and fuse the two mixtures together.

White Enamel.

For white enamel all ingredients must be perfectly free from foreign admixture as any impurities will interfere with the color. Washed antimony 1 part, fine glass, perfectly free from lead, 3 parts; mix, melt and pour the fused mass into clean water; dry and repeat as before until a perfectly uniform color is obtained. When well managed the color will rival the opal.

Violet Enamel.

Saline or alkaline frits or fluxes colored with small quantities of peroxide of manganese. The color depends on the complete oxidization of the metal and care should therefore be taken not to give contact with any substance in the flux that will abstract the oxygen.

Rose Colored Enamel.

Purple enamel or its constituents, 3 parts, flux 90 parts. Mix and add silver leaf or oxide of silver one part or less, according to shade desired.

Purple Enamel.

1. Flux, colored with oxide of gold, purple precipitate of cassius, or peroxide of manganese.

2. Sulphur, niter, vitriol, antimony, oxide of tin, of each 1 pound, red lead 60 pounds, mix, fuse, cool and powder, add rose copper, 19 oz., zaffre 1 oz., crocus martis, 1½ oz., borax, 3 oz., 1 lb. of a compound, formed of gold, silver and mercury. Fuse, stirring constantly with a copper rod, place in crucibles and place in reverberating furnace 24 hours.

Olive Enamel.

Good blue enamel 2 parts, black and yellow enamel, of 1 part each. Mix.

Blue Enamel.

1. Thoroughly powder and mix 4 parts of black oxide of cobalt, 13 of saltpetre, and 9 of flint. Fuse thoroughly over a charcoal or coke fire, pulverize, wash in cold water, and triturate 1 part of this powder with one of flux 5, given above.

2. Fuse together 1 part of black oxide of cobalt and 1 of borax. Melt over a good fire 2 parts of this, ½ part of minium and 10 of blue pot metal glass.

Yellow Enamel.

1. Mix in a stone mortar 8 parts of minium, 1 of white oxide of tin and 1 of antimony oxide. Bring to a red heat in a crucible, allow to cool off and mix 1 part of this with 4½ parts of flux 4 to a paste with water.

2. White oxide of antimony, alum and sal ammoniac, each 1 part, pure carbonate of lead 1 to 3 parts, according to shade desired. Powder, mix and expose to heat sufficient to fuse the sal ammoniac very bright.

3. Pure oxide of silver added to the metallic fluxes. If a thin film of oxide of silver be spread over the surface of the enamel to be colored, exposed to a moderate heat, withdrawn and the film of reduced silver on the surface removed, the under part will be found a fine yellow.

Orange Enamel.

1. Red lead 12 parts, red sulphate of iron and oxide of antimony, each one part, flint powder 3 parts. Calcine, powder, and melt with flux 50 parts.

2. Red lead 12 parts, oxide of antimony 4 parts, flint powder 3 parts, red sulphate of iron 1 part. Calcine, then add flux 5 parts to every 5 parts of this mixture.

3. Thoroughly mix 12 parts of minium, 4 of antimony oxide, 3 of pulverized flint and 1 of red sulphate of iron, and heat without fusing. Triturate with water 1 part of this and 2½ of flux 7.

Enamels for Watch Dials.

The dials are prepared with a backingof sheet copper having raised edges to receive the enamel in powder, which is fused. After cooling, the lettering and figuring are printed on the plate with soft black enamel by transferring. The dial is again placed in a muffle to fuse the enamel of the lettering or figuring. The enamel used is composed of white lead, arsenic, flint glass, saltpetre, borax and ground flint reduced to powder, fused and formed into cakes.

Phosphorescent Enamel.

Commercial phosphorescent paint in powder is intimately mixed with ⅔ of its weight of very finely pulverized fluor spar or cryolite and one-fifth of calcium borate. The mixture is made into a paste with water, and applied in a uniform layer to the articles to be enameled by means of a brush. They are then burnt in the usual manner.

Fine Black Enamel.

Triturate with water 1 part of black oxide of copper and 2 of flux 4.

Niello.

A metallic enamel composed of 48 parts of flowers of sulphur, 9 of pure copper, 9 of pure lead, 4 of fine silver and two of borax. First melt the silver, then add the copper and then the lead, stirring with a charcoal stick. Prepare a crucible, into which place the sulphur and upon the top of which pour the molten metal, and continue to fuse a few minutes. Pour the mass over brushwood into water to form into granules. Collect the granules, dry in the air and pulverize in a mortar. Niello is used as follows: the design is engraved on the metal to be ornamented and the powder is mixed with spirit of sal ammoniac to a paste, and applied by rubbing into the lines in the metal which has previously been heated. The superfluous paste is renewed by filing and the surface stoned and polished.

To Take Enamel Out of Work.

Take fluor spar, and pound it fine; use enough of the powder to cover well the article; then take a lead cup and pour on sufficient vitriol to make a thin liquid. Boil the article in this and the enamel will be removed. Be very careful and let the fumes pass up chimney, as they are injurious.

BRONZING, STAINING AND COLORING OF METALS.

Green Bronze for Brass.

1. Mix 80 parts strong vinegar, 1 of mineral green, 1 of red umber, 1 of sal ammoniac, 1 of gum Arabic, 1 of green vitriol, and add 4 of Avignon berries (fruit of *Rhamus Infectorius*). Boil the mixture and strain when cold. Cleanse the articles with weak nitric acid, rinse and apply the fluid with a brush. If not dark enough heat the article to about 200°, and then give a coat of lamp black mixed with alcohol. Finish with lacquer or spirit varnish.

2. Dissolve 8½ ounces of copper in 1 ounce nitric acid; add 10½ fluid drams vinegar, 3½ drams sal-ammoniac, 6¾ drams aqua ammonia. Put the liquid in a loosely corked bottle, and stand in a warm place for a few days. After applying it to the articles, dry by gentle heat and finish with linseed oil varnish, also dried by heat.

Violet on Brass.

Heat the metal to about the boiling point, plunge it in a solution of chloride of antimony for an instant and rub with a stick covered with cotton.

Orange on Brass.

Polish the article and plunge it for a few minutes in a warm neutral solution of copper acetate.

Green on Brass.

1. Wash the article with dilute acetic acid and expose to the fumes of strong aqua ammonia, Repeat until the desired shade is obtained.

2. Immerse in a solution of 1 part perchloride of iron in 2 parts of water until the desired shade is assumed.

3. Boil in a strong solution of nitrate of copper. Wash, dry, burnish and lacquer all the above.

Steel Gray on Brass.

Antimoniac sulphide and fine iron filings, 1 part of each; hydrochloric acid 3 parts; water 4 parts.

Blue on Brass or Copper.

Cleanse the article thoroughly by boiling in potash lye and treat it with white wine vinegar; wipe and dry the surface thoroughly and rub with a linen rag, moistened with hydrochloric acid; let it stand for a quarter of an hour, then heat the article over a sand bottle until it assumes the desired shade.

Dead Black on Brass.

Mix in a glass bottle 2 parts of hydrochloric acid and 1 part of nitric acid; add 35 grains of platinum wire or foil, and place on a sand bath until the platinum is dissolved. Apply by dipping, or with a brush. This is the finish used on optical instruments.

Black Bronze for Brass.

Brush the brass with a dilute solution of nitrate of mercury, and then several times with a solution of liver of sulphur.

To Frost Watch Caps and Plates.

Take 2½ parts nitric acid, and 2 parts muriatic acid, full strength. Dip in the articles for a few seconds, rinse in clear water, scratchbrush with a circular motion, then gild.

To Frost Watch Plates.

Watch plates are frosted by means of fine brass wire scratch brushes fixed in a lathe, and made to revolve at great speed, the end of the wire brushes striking the plate producing a beautiful appearance; or, sink that part of the movement to be frosted for a short time into a mixture of nitric acid, muriatic acid and table salt, 1 ounce of each. On removing from the acid, place it in a shallow vessel containing enough sour beer to nearly cover it, then with a fine scratch brush scour thoroughly, letting it remain under the beer during the operation. Then wash off, first in pure water and then in alcohol. Gild or silver in accordance with any of the well known methods.

To Frost Silver.

1. Silver goods may be frosted and whitened by preing a pickle of sulphuric acid 1 dram, water 4 ounces; heat and immerse the silver articles until frosted as desired; then wash off clean, and dry with a soft linen cloth, or in fine clean sawdust. For whitening only, a smaller quantity of acid may be employed.

2. The article has to be carefully annealed either in a charcoal fire or with a blow pipe before a gas flame, which will oxidize the alloy on the surface, and also destroy all dirt and greasy substances adhering to it, and then boiled in a copper pan containing a solution of dilute sulphuric acid—of 1 part of acid to about 30 parts of water. The article is then placed in a vessel containing clean water, and scratch-brushed or scoured with fine sand, after which

the annealing and boiling-out is repeated, which will in almost all cases be sufficient to produce the desired result. If a very delicate dead surface such as watch dials, etc., is required, the article is, before the second annealing, covered with a pasty solution of potash and water, and immediately after the annealing plunged in clean water, and then boiled out in either sulphuric acid solution, or a solution of 1 part cream tartar and 2 parts common salt to about 30 parts of water. If the article is of a low quality of silver, it is well to add some silver solution, such as is used for silvering, to the second boiling-out solution. If the article is very inferior silver, the finishing will have to be given by immersing it in contact with a strip of zinc in a silver solution.

To Frost Polished Silver.

Dissolve ½ oz. cyanide of potasium in 4 oz. water, apply with a brush, holding the article with lancewood or boxwood pliers. Cyanides are very poisonous and should be used with care.

Silvering Copper and Brass.

Mix 3 parts of chloride of silver with 20 parts finely pulverized cream of tartar, and 15 parts culinary salt. Add water in sufficient quantity, and stir until the mixture forms a paste, with which cover the surface to be silvered by means of blotting paper. The surface is then rubbed with a rag and powdered lime, washed, and rubbed with a piece of soft cloth. The deposited film is extremely thin.

Silver Plating.

For rapid silver plating, prepare a powder of 3 parts of chloride of silver, 20 parts carefully pulverized cream of tartar, and 15 parts pulverized cooking salt; mix it into a thin paste with water, and rub it upon the well-cleaned metallic surface with blotting paper. After you are certain that all parts of the article have been touched alike, rub it with very fine chalk powder or dust upon wadding or other soft cloth. Wash with clean water and dry with a cloth.

Silver-Plating Fluid.

Dissolve 1 oz. nitrate of silver, in crystals, in 12 oz. soft water; then dissolve in the water 2 oz. cyanide of potash, shake the whole together, and let it stand until it becomes clear. Have ready some half-ounce vials, and fill half full with Paris white, or fine whiting, and then fill up the bottles with the liquid, and it is ready for use. The whiting does not increase the coating power, it only helps to clean the article, and save the silver fluid, by half filling the bottles.

Simple Method for Silver Plating.

The process consists in exposing the article, which has previously been well cleansed with a potash solution and dilute hydrochloric acid, to the operation of a silver bath, which is prepared in the following manner: Form a solution of 32 grams (1 oz., 13.8 grains) nitrate of silver, 20 grams silver (12 dwts., 20.6 grains) in 60 (1 oz., 18 dwts., 13.9 grains) grams nitric acid. The silver is precipitated as silver oxide with a solution of 20 grams solid caustic potash in 50 grams (1 oz., 12 dwts.; 3.6 grains) distilled water, carefully washed, and the precipitate taken up by a solution of 100 grams (3 oz., 4 dwts., 7.2 grains) cyanide of potassium in 500 grams distilled water. The fluid, distilled through paper, is finally diluted with distilled water, to 2 liters (4¼ pints). The thus prepared silver bath is gently warmed in the water bath, and the article to be silver plated laid in it and kept in motion for a few minutes, and after taking out it is dried in sawdust, and then polished with Vienna chalk for giving luster.

Silvering Receipt.

Care must be taken that the pieces which are dipped in the metal bath be treated before in the ordinary manner in a potash solution and dilute hydrochloric acid. The silver bath is made with a solution of 4 ounces lunar caustic (equal to a solution 2½ ounces silver in 7½ ounces nitric acid); the silver of this solution is precipitated as oxide of silver by the addition of a solution of 2½ ounces of caustic potash in 6½ ounces distilled water; and the precipitate, after being washed, is added to a solution of 12½ ounces of cyanide of potassium in one quart of water. This solution is then filtered and water added to bring it to 4 quarts. In this solution, which is heated on the water bath, the pieces that are to to be silvered are left for a few minutes. Being agitated, they are taken out, and put to dry in fine sawdust and then polished.

Silvering Small Iron Articles.

The small iron articles are suspended in dilute sulphuric acid until the iron shows a bright clean surface. After rinsing in pure water, they are placed in a bath of a mixed solution of sulphate of zinc, sulphate of copper and cyanide of potassium, and there remain until they receive a bright coating of brass. Lastly they are transferred to a bath of nitrate of silver, cyanide of potassium, and sulphate of soda, in which they quickly receive a coating of silver.

Cold Silvering of Metal.

Mix 1 part of chloride of silver with 3 parts of pearl-ash, 1½ parts common salt, and 1 part whiting; and well rub the mixture on the surface of brass or copper previously well cleaned, by means of a piece of soft leather, or a cork moistened with water and dipped in the powder. When properly silvered, the metal should be well washed in hot water, slightly alkalized, then wiped dry.

Gold Tinge to Silver.

A bright gold tinge may be given to silver by steeping it for a suitable length of time in a weak solution of sulphuric acid and water, strongly impregnated with iron rust.

Imitation of Antique Silver.

The article is dipped in a bath of water containing about 10 per cent. of sulphide of ammonium, and then scratch-brushed with a brush made of glass threads or bristles. When afterward burnished with an agate tool its surface becomes a beautiful dark brown color.

Oxidizing Silverware.

Sal-ammoniac, 2 parts; sulphate of copper, 2 parts; saltpeter, 1 part. Reduce these ingredients to a fine powder, and dissolve it in a little acetic acid. If the article is to be entirely oxidized, it may be dipped for a short time in the boiling mixture; if only in parts, it may be applied with a camel-hair pencil, the article and the mixture both being warmed before using.

Oxidizing Silver.

There are two distinct shades in use, one produced by chloride, which has a brownish tint, and the other by sulphur, which has a bluish-black tint. To produce the former it is only necessary to work the article with a solution of sal-ammoniac; a much more beautiful tint, however, may obtained by employing a solution composed of equal parts of sulphate of copper and sal-ammoniac in vinegar. The fine black tint may be produced by a slightly warm solution of sulphate of potassium or sodium.

Oxidizing Silver Brown-black.

Place the articles in a solution of equal parts of sal-ammoniac and blue vitriol in vinegar until the desired shade is obtained. Then rinse, dry and polish.

Oxidizing Silver Blue-black.

Place the articles in a solution of liver of sulphur diluted with spirits of sal-ammoniac. Allow to remain until sufficiently discolored, then wash, dry and polish.

Pink Tint on Silver.

Dip the cleaned article for a few seconds in a hot solution of chloride of copper; then rinse and dry, or dip in 90 per cent. alcohol and ignite the alcohol.

Dead White on Silver Articles.

Heat the article to a cherry-red or a dull red heat, and allow it to cool, then place it in a pickle of 5 parts sulphuric acid to 100 parts of water, and allow it to remain for an hour or two. If the surface is not right, rinse in cold water, and repeat the heating and pickling operation as before. This removes the copper from the surface of the article, leaving pure silver on the surface. When sufficiently whitened, remove from the pickle, well rinse in pure hot water, and place in warm boxwood sawdust.

To Whiten Silver Watch Dials.

Flatten a piece of charcoal by rubbing it on a flat stone; on this place the dial, face upward; apply a gentle heat carefully with a blowpipe, allowing the flame to play all over the surface of the dial without touching it, so as to thoroughly heat without warping the dial. Then pickle and rinse, using acid enough to make the water very tart, and immersing but for a few seconds. Silver dials may also be annealed by heating them red hot on a flat piece of copper over a clear fire.

Gold Yellow for Brass.

A gold like appearance may be given to brass by the use of a fluid prepared by boiling for about 15 minutes, 4 parts caustic soda, 4 parts milk sugar, and 100 parts water, after which 4 parts of a concentrated solution of sulphate of copper is added with constant stirring. The mixture is then cooled to 79 degrees C., and the previously well cleaned articles are for a short time laid into it. When left in it for some time they will first assume a blueish and then a rainbow color.

Coloring Copper.

To produce a dark-brown color upon copper, take the white of an egg, beat it into froth, add a little boiled or rain water, and add to this mixture *caput mortuum* (red oxide of iron) color; rub them well together in a mortar, and sufficiently thick until the color covers, and may be applied. The copper articles are to be pickled and simply washed; no sand must be used, else the color adheres badly. The latter is next applied with a brush until it covers the surface; it is then dried by a fire, the article is gently rubbed with a soft rag and *caput mortuum* powder, and finally hammered with a hammer with polished face.

Transparent Blue for Steel.

Damar varnish, 1 pint; finely pulverised Prussian blue, 1 dram; mix thoroughly. Makes a splendid appearance. Excellent for blueing hands.

To Blue Steel.

In order to blue steel pieces evenly, the following will give satisfactory results; first blue the object without any special regard to uniformity of color. If it proves to be imperfect, take a piece of deadwood that does not crumble too easily, or of clean pith, and whiten the surface with rouge without letting it be too dry. Small pieces thus prepared, if cleaned and blued with care, will assume a very uniform tint.

To Bronze Steel.

Methylated spirit, 1 pint; gum shellac, 4 ounces; gum benzoine, ½ ounce. Set the bottle in a warm place, with occasional agitation. When dissolved, decant the clear part for fine work, and strain the dregs through muslin. Now take 4 ounces powdered bronze green, varying the color with yellow ochre, red ochre and lamp black, as may be desired. Mix the bronze powder with the above varnish in quantities to suit, and apply to the work, after previously cleansing and warming the articles, giving them a second coat, and touching off with gold powder, if required, previous to varnishing.

To Blue Screws Evenly.

Take an old watch barrel and drill as many holes into its head as you desire to blue screws at a time. Fill it about one-fourth full of brass or iron filings, put in the head, and then fit a wire long enough to bend over for a handle into the arbor holes—head of barrel upward. Brighten the heads of your screws, set them point downward into the holes already drilled, and expose the bottom of the barrel to your lamp, until the screws assume the color you wish.

Aniline Bronzing Fluid.

A bronzing fluid which is said to be very brilliant, and applicable to all metals, as well as to other substances, is prepared as follows: Take 10 parts of aniline red and 5 parts of aniline purple, and dissolve in 100 parts of 95 per cent. alcohol, accelerating the solution by placing the vessel in a sand or water bath. Solution having been effected, add 5 parts of benzoic acid, and boil for from 5 to 10 minutes, until the greenish color of the mixture has been converted into a fine, light-colored bronze, which is applied with a brush and dries easily.

Antique Bronzes.

One can give bronze the green stain of verdigris by covering the spots to be discolored with ground horse-radish saturated with vinegar, and keeping the horse-radish wet until the stain has become fixed. This will require some days; for though the discoloration will show after a few hours, it will be superficial and vanish by wiping. Three or four days will, however, turn your bronze into an antique, so far as the mockery of age can make it old.

To Color Soft Solder.

The following is a method for coloring soft solder so that when it is used for uniting brass the colors may be about the same: First prepare a saturated solution of sulphate of copper—blue stone—in water, and apply some of this on the end of a stick to the solder. On touching it then with an iron or steel wire it becomes coppered; and by repeating the experiment the deposit of copper may be made thicker and darker. To give the solder a yellow color, mix one part of a saturated solution of sulphate of zinc with two of sulphate of copper; apply this to the coppered spot and rub with a zinc rod. The color can be still further improved by applying gilt powder and polishing. On gold jewelry or colored gold the solder is first coppered as above, then a thin coat of gum or isinglass solution is laid on and bronze powder dusted over it, making a surface which can be polished smooth and brilliant after the gum is dry.

Silvering Tincture.

Experiments have shown the following receipts for a silvering tincture to be excellent. Prepare the following solutions:

A. Two parts of burned lime, 5 parts of grape sugar, 2 of tartaric acid, 650 of water. The solution is filtered and put in bottles; should be entirely filled and well corked.

B. Dissolve 20 parts of nitrate of silver in 20 of aqua ammonia and then add 650 of water.

Just before the tincture is to be used mix solutions A and B together; shake well and filter. Metals and dry vegetable substances, such as wood tissues, horn buttons, ivory, etc., can be silvered with this fluid.

To Bronze Medals, etc.

Powder and mix 1 pound each of verdigris and salammoniac; take a portion of this about as large as a hen's egg and boil in a copper pan with about 5 pints of water for 20 minutes. Let it settle and pour off the water. Place the medals in a copper pan, resting them on pieces of wood or glass, so they do not touch each other or the copper; pour the fluid upon them and boil until the desired color is obtained.

Chinese Brown Bronze.

Powder and mix thoroughly 2 parts crystallized verdigris, 2 of cinnabar, 2 of sal-ammoniac, 2 of horn shavings and five of alum. Moisten with water or alcohol and rub into a paste. Cleanse the articles thoroughly, polish with ashes and vinegar, apply the paste with a brush, heat over a coal fire and wash the coating off. Repeat until the desired shade is obtained. Addition of blue vitriol gives a chestnut brown, while borax gives a yellowish shade.

Antique Green.

Dissolve 1 part sal-ammoniac, 3 of powdered tartar, 3 of common salt, in 12 of boiling water. Then add 8 parts cupric nitrate, and coat the articles with the liquid.

To Cover Spots on Gold or Plated Articles.

The following recipe will be found to answer well in removing or covering over spots on gold or plated articles where the plate is worn off. Dissolve twenty-four grains of fine coin gold in one-half ounce of nitro-muriatic acid, and then absorb the acid with a clean blotting paper. When the paper is thoroughly dry burn it and pulverize the ashes, which rub on the spots with chamois skin, moistened with water. The spots should first be thoroughly cleaned.

Acid-Coloring Small Articles.

For acid-coloring on small articles, a very good plan is to place them on a lump of charcoal, and make them red hot under the blow-pipe flame, and then to throw them into a pickle composed of about 35 drops strong sulphuric acid to 1 ounce of water, allowing the article to remain therein until the color is sufficiently developed; washing the article in warm water in which a little potash has been dissolved, using a brush, and finally rinsing and drying in boxwood sawdust, completes the operation.

To Whiten Iron.

Take ammoniacal salt in powder and mix it with an equal quantity of quick silver. Dissolve in cold water and mix well. Immerse the red heated metal in this bath and it will become as white and beautiful as silver. Be careful and do not burn the article by overheating.

LACQUERS AND VARNISHES.

In lacquering metals of all kinds, be sure that all oil and grease are removed from the surface; the work should not be handled with the fingers, but should be held with a spring tongs or with a clean cloth. The work should be heated so hot that the brush will smoke on being applied, but not so hot as to burn the lacquer. See that the end of the brush is perfectly even; if not trim it. Use the extreme end of the brush and very little lacquer, as it is better to apply two thin coats than one thick one. If the lacquer be too thick, it should be thinned with a proper medium, as alcohol, turpentine, etc., and if too thin, evaporate by placing on the stove. Where articles are lacquered in large quantities they may be dipped, in which case they should be immersed by means of a wire into a bath composed of equal parts of nitric and sulphuric acids, removed, rinsed thoroughly in cold water, dipped into hot water, then in alcohol and then dipped momentarily into lacquer, shaken to remove all superfluous lacquer and laid on a warm metal plate until dry. Lacquer for dipping should be considerably thinner than that used with a brush.

Lacquer For Silverware.

Coat the article with a fine brush with collodion, which has previously been diluted pretty strongly with alcohol. This coating dries at once and forms a very thin, transparent and invisible protection which shields the silver completely, and, if necessary, may be washed off with hot water. This process is much employed in English silver stores.

Lacquers For Brass.

1. Dragon's blood 40 grains; seed lac 6 ounces; amber and copal, triturated in a mortar, 2 ounces; oriental saffron 36 grains; alcohol 40 ounces; extract of red sanders ½ dram; coarsely powdered glass 4 ounces.

2. Gamboge, seed lac, annatto, dragon's blood, each 1 ounce; 2½ pints alcohol, ¼ ounce saffron.

Gold Lacquer for Brass.

Twenty-four grains extract red sanders wood in water, 60 grains dragon's blood, 2 ounces amber, 6 ounces seed lac, 2 ounces gamboge, 36 grains oriental saffron, 36 ounces pure alcohol; 4 ounces powdered glass. The amber, gamboge, glass, dragon's blood and lac should be thoroughly pounded together. Infuse the saffron and the sanders wood extract in the alcohol for 24 hours. Pour this over the other ingredients and strain.

Lacquer for Brass.

Coat it with the following varnish: 1 part white shellac and 5 alcohol; 1 shellac, 1 mastic, 7 alcohol; or, 2 sandarac, 8 shellac, 1 Venetian turpentine, 50 alcohol; or, 12 parts sandarac, 6 mastic, 2 elemi, 1 Venetian turpentine, 64 alcohol. Clean the article well, do not touch with your hands, and warm to about 75° C.

Transparent Lacquer.

Dissolve 2 parts camphor and 30 parts copal gum in 30 parts of oil of lavender and 120 parts of oil of turpentine.

Fine Pale Lacquer.

Saffron 2 drms., white shellac 1 oz., turmeric 1 drm., Gamboge 1 drm., alcohol 1 pint.

Simple Pale Lacquer.

Dissolve 1 oz. white shellac in a pint of alcohol.

Green Lacquer.

To 1 pint simple lacquer add 4 drams turmeric, 1 dram of Gamboge.

Red Lacquer.

To 1 pint simple lacquer add 32 drams annatto and 8 drams of dragon's blood.

Gold Lacquer.

Turmeric 16 drms., shellac 3 oz., saffron 2 drms., annatto 2 drms., alcohol 1 pint.

Amber Lacquer.

Oil of turpentine 12 parts, amber 4 parts, Venice turpentine 1 part, elemi 1 part.

Gold Lacquer on Iron.

Dissolve 3 oz. finely powdered shellac in 1¾ pints of alcohol. Filter through linen and rub in a sufficient quantity of dutch gold to give a lustrous color. Polish and heat the iron, brush with vinegar and then apply the color with a brush, when dry, varnish with copal varnish.

Lacquer for Dials.

A handsome varnish for dials of clocks, watches, etc., may be prepared by dissolving bleached shellac in the purest alcohol. It offers the same resistance to atmospheric influence that common shellac does, when used as a coating on brass. The manner of applying it is easily learned.

Lacquer for Steel.

After having cleaned the iron or steel article, anoint it with a solution of wax in benzine, using a fine camel's hair brush. By this treatment, articles exposed to acid vapors, may be protected against rusting. Another coating may be made if the steel or iron is covered with a layer of a mixture obtained by boiling sulphur with turpentine oil; this evaporates and leaves the sulphur upon the surface as pure sulphur, which again combines with the metal and forms sulphuret of iron, by heating the articles, if small, over a gas or alcohol flame,

Black Lacquer for Iron and Steel.

Boil 1 part of sulphur with 10 parts oil of turpentine. Cover the article with a very thin coat and hold over the flame of an alcohol lamp, until the black polish makes its appearance.

Lacquer for Metals.

Melt one part by weight of best wax paraffine, and when sufficiently cooled, add three parts of petroleum. Mix well together, and apply to the polished article by means of a soft brush. The protecting film need only be very thin, wherefore not too much should be applied.

Lacquer for Gypsum Figures.

Three parts caustic potash are dissolved in 36 parts hot water, 9 parts stearic acid are added, and the obtained soap paste is diluted with the same quantity of water and 95 per cent. alcohol. The warm solution is applied upon the warm gypsum cast, and this, after a few hours, is repeated with a wet sponge. The casting becomes still handsomer if, in place of potash, a corresponding quantity of ammonia is used. Old casts are first cleaned with a 3 per cent. caustic potash solution.

CLEANING, PICKLING AND POLISHING.

Polishing Agents.

Various polishing agents are used by watchmakers, jewelers, gold and silversmiths, a few of which are here described. Where the article will admit of it, the best results are obtained by polishing in the lathe. For this purpose the watchmaker should not use his regular lathe, but should have for the purpose what is known as a polishing lathe, fitted with its various attachments in the shape of scratch brushes, buffs, etc.

Polishing Powder for Gold.

1. White lead, 43 parts; chalk, 174 parts; carbonate of magnesia, 17 parts; alumina, 43 parts; silica, very finely powdered, 26 parts; ferric oxide, 17 parts. This is an excellent powder and is much used in finishing new work by goldsmiths.

2. Mix together 4.3 parts of alumina, 17.4 of chalk, 4.3 of carbonate of lead, 1.7 of carbonate of magnesia, and 1.7 of rouge.

3. An excellent polishing powder for gold and silver consists of burnt and finely pulverized rock alum, 5 parts, and powdered chalk 1 part. Mix and apply with a dry brush.

Restoring the Color to a Gold or Gilt Dial.

Dip it for a few seconds in the following mixture: Half an ounce of cyanide of potassium, is dissolved in a quart of hot water, and 2 ounces of strong ammonia, mixed with half an ounce spirits of wine, are added to the solution. On removal from the bath, the dial is immediately immersed in warm water; then brush with soap, rinse, and dry in hot boxwood dust. Or it may be simply immersed in dilute nitric acid, but in this case any painted figures will be entirely destroyed.

Removing Spots on Gilding.

Boil common alum in soft, pure water and immerse the article in the solution, or rub the spot with it and dry with sawdust.

Cleaning Electro-Plate.

The tarnish can be removed by dipping the article from one to fifteen minutes, in a pickle of the following composition: Rain water, 2 gallons, and potassa cyanuret ½ pound; dissolve together, and fill into a stone jug or jar, and close tightly. The article after having been immersed, must be taken out and thoroughly rinsed in several waters, then dried with fine, clean sawdust. Tarnished jewelry can speedily be restored by this process; but be careful to thoroughly remove the alkali, otherwise it will corrode the goods.

Cleaning Gold Tarnished in Soldering.

It is usually cleaned in dilute sulphuric acid. The pickle is made in about the proportion of one-eighth of an ounce of acid to one ounce of rain water.

Cleaning Mat Gold.

Take 80 gr. chloride of lime, 80 gr. bicarbonate of soda, and 20 gr. table salt; pour over this about 3 quarts distilled water, and put in bottles, to be kept well corked. For use, lay the dirty articles in a dish, pour over them the well shaken fluid, let it submerge them, leave them in it for a short time, and in extra cases, when very dirty, warm them a little. Next wash the articles, rinse them in alcohol, dry them in sawdust, and they will appear like new. The fluid is of no further use.

Cleaning Watch Chains.

Gold or silver watch chains can be cleaned in a very excellent manner—no matter whether they be mat or lustrous—by laying them for a few seconds in pure aqua ammoniac; they are then rinsed in alcohol, and finally shaken in clean sawdust, free from sand. Imitation and plated chains are first cleaned in benzine, they are then rinsed in benzine, and afterward shaken in dry sawdust. Ordinary chains are first to be dipped in the following pickle: Pure nitric acid is mixed with concentrated sulphuric acid, at the rate of 10 parts of the former and two parts of the latter; a little cooking salt is mixed with this. The chains are boiled up in this mixture; they are then rinsed several times with water, finally in alcohol, and dried in sawdust.

Silver Soaps.

1. Cut in small pieces 2 lb. cocoanut oil soap; dissolve in sufficient water to form a thick jelly; add 2 lb. fine rouge by stirring until thoroughly homogeneous, and put in boxes.

2. Dissolve 14 oz. Marseilles soap in 2 quarts of water, add 7 oz. finest French chalk; if color is desired, add a little fuschine; bottle for use. Apply with a woolen rag.

3. Saponify 10 lb. of cocoanut oil with pure caustic soda in the usual manner and boil to a clean jelly, then add 2 lb. tripoli, 1 lb. alum, 1 lb. cream tartar, 1 lb. white lead. All the ingredients should be very finely powdered

and intimately mixed before stirring in the soap. Pour the mixture into tin moulds and it will quickly solidify. To use, moisten the article with lukewarm water and apply the soap with a rag.

4. Hard soap, 8 oz., turpentine 1½ oz., water 4 oz, boil until perfectly dissolved and add aqua ammonia 3 oz.

5. Dissolve 10 parts castile soap in 10 parts water; remove from the fire and stir in 30 parts fine whiting.

6. Dissolve 10 parts castile soap in 10 parts of water; remove from the fire and stir in 10 parts tripoli, 5 parts rouge, 15 parts French chalk. The powders should be fine, and intimately mixed before adding.

Polishing Powder for Silver.

Mix intimately, 4 parts of finest washed pipe clay and 1 part of pure tartar.

Cleaning Silverware.

Hyposulphate of soda is the simplest and most effective cleansing material for silverware; it operates quickly and is cheap. A rag or a brush moistened with the saturated solution of the salt cleans, without the use of cleaning powder, strongly oxidized silver surfaces within a few seconds.

Cleaning Silver Tarnished in Soldering.

Expose to a uniform heat, allow it to cool, and then boil in strong alum water; or, immerse for a considerable length of time in a liquid made of ½ oz. of cyanuret of potassia to one pint of rain water, and then brush off with prepared chalk.

Cleaning Silver Filigree Work.

Anneal your work over a Bunsen flame or with a blowpipe, then let grow cold (and this is the secret of success), and then put in a pickle of sulphuric acid and water, not more than five drops to one ounce of water, and let your work remain in it for one hour. If not to satisfaction, repeat the process.

Brass Polishes.

1. Rottenstone 4 oz., oxalic acid, powdered, 1 oz.; sweet oil, 1½ oz.; turpentine to make a paste; apply with soft leather.

2. Equal parts of sulphur and chalk, made into a paste with vinegar. Allow to dry on the article and clean with a chamois or brush.

3. Dip the brass in a mixture of 1 oz. alum, 1 pint lye and polish with tripoli on a chàmois. This gives a brilliant luster.

Magic Polish for Brass.

Add to sulphuric acid half its bulk of bichromate of potash; dilute with an equal weight of water, and apply well to the brass; rinse it well immediately in water, wipe dry, and polish with pulverized rotten stone.

Polishing Paste for Brass.

Dissolve 15 parts of oxalic acid in 120 parts of boiling water and add 500 parts of pumice powder, 7 of oil of turpentine, 60 of soft soap, and 65 of fat oil. The polishing agent is usually mixed with oil, alcohol or water, to prevent scattering, and is then applied to the polishing tool in the shape of cloth and leather buffs, polishing files, etc. Either the work or the tool should revolve with great velocity in order to secure good results. Many articles are brought to a high degree of polish by the use of the burnisher, after subjecting them to the action of the ordinary polishing agents.

To Clean Brass.

The method prescribed for cleaning brass, and in use in all the U. S. arsenals, is claimed to be the best in the world. The plan is to make a mixture of one part common nitric acid and one-half part sulphuric acid in a stone jar, having also ready a pail of fresh water and a box of sawdust. The articles to be treated are dipped into the acid, and then thrown into the water, and finally rubbed with sawdust. This immediately changes them to a brilliant color. If the brass has become greasy, it is first dipped in a strong solution of potash and soda in warm water; this cuts the grease, so that the acid has free power to act.

To Polish a Watch Wheel.

It can be done nicely in the following manner: get a cork flat on the top, and put into a vise; on it place the wheel, as far as the pinion will allow; then take a bluestone and water, and grind the wheel smooth and flat, all the time revolving it with the left hand; wash it, and put it in a box with some slaked powdered lime. This is done simply for the purpose of drying it, and preventing the pinion from getting stained or rusty. Brush it out nice and clean, put another cork, clean and flat, in the vise, and pound some crocus on a stake. Some workmen add a little rouge, but this is simply a matter of taste. Take a slip of tin, about the size of a watchmaker's file, only thicker, file the end of one side flat and smooth, charge it with a little of the crocus, and polish the wheel, all the time rotating it with your left hand; do not cease until both wheel and tin polisher are almost dry, so that you can see the polish, when, if to your satisfaction, clean the wheel off with a piece of soft bread, and brush it out. Should it be scratched, bread them off, clean off the tin, and take a new supply of crocus. Cleanliness in this manipulation is of the greatest importance, for if there should be any grit about the crocus, polisher, or the fingers of the workman, the work will be full of scratches. This applies simply to bar wheels.

To Polish Jewel Settings.

A very good way to polish jewel settings to American watches on brass or gold, is as follows: First turn the setting down to the right thickness, or nearly so, and then grind down to a gray on a ground glass slab with rotten stone and oil; then clean off the oily rotten stone and polish on a boxwood lap with diamantine and oil, which gives a nice gloss. It will also give a nice gloss on steel, only use oil stone to gray steel with, instead of rotten stone. The operator should be particular to clean off all the graying powder in each case before using the box-wood lap, and be sure to keep the lap in a place free from grit or dust when not in use; brass watch wheels can be finished in the same way as the jewel settings by the same process.

Friction Polish on Steel.

1. After turning as smooth as possible, dress with rouge on a bar of pure tin, using considerable pressure and very little rouge, revolving in a lathe at high speed.

2. After turning, polish with rouge and use a hardened steel or agate burnisher with a little oil.

Polishing Steel.

1. Take crocus of tin oxide, and graduate it in the same way as preparing diamond dust, and apply it to the steel by means of a piece of soft iron or bell metal, made in proper form, and prepared with flour of emery, same as for pivot burnishers; use the coarsest of the crocus first, and finish off with the finest. To iron or soft steel

a better finish may be given by burnishing than can be imparted by the use of polishing powder of any kind whatever. The German method of polishing steel is performed by the use of crocus on a buff wheel. Nothing can exceed the surpassing beauty imparted to steel or even cast iron by this process.

2. If the steel is of moderately good temper, use a zinc polisher with diamantine; a tin polisher is better for soft steel. The diamantine should be mixed on glass, using a beater, also of glass, with very little watch oil. Diamantine mixed with oil becomes gummy, and quite unfit in a day or two, and turns black, if brought into contact with metal, in mixing.

3. To polish such parts as rollers and collets, first get a flat surface, by rubbing with fine emery on a glass plate or a bell-metal block, and afterward finish off on a zinc block with diamantine; but for levers, you must use a long flat bell-metal or zinc polisher, and press the lever into a piece of soft wood (willow is the best) in the vise, moving the polisher instead of the work. For large articles, such as indexes or repeater racks, which are not solid, and spring, it will be found best to wax them on to a small brass block and polish them underhand, in the same manner as rollers.

4. Mix 1 pound of fine colcothar in 5 pounds common yellow vaseline. Apply with a rag or wash leather and rub clean.

Removing Rust from Steel.

For cleaning purposes, etc., kerosene oil or benzine are probably the best things known. When articles have become pitted by rust, however, these can only be removed by mechanical means, such as scouring with fine powder, or flour of emery and oil, or with very fine emery paper. To prevent steel from rusting, rub it with a mixture of lime and oil, or with mercurial ointment, either of which will be found valuable.

Removing Rust from Pinions.

The best way to remove rust from pinions is to scour them up with oilstone dust and oil, till a smooth surface is obtained, then polish with crocus. Care must be taken not to grind the leaves off any more than is necessary, or the proper shape may be destroyed. Some workmen soak the rusted parts in a solution of cyanide of potassium or other solvent of oxide of iron, but the use of such means cannot be approved of. The way described is as good as any, and is safe. If the pinions are very badly rusted they should be rejected and others put in, as they will be out of shape when finished off smooth, and would not perform well in the watch.

Cleaning Files.

Let them lie in benzine until the metals, grease, etc., have been thoroughly soaked and then remove them by scratch brushing.

Removing Rust from Nickel.

1. Cover the stains with olive or cocoanut oil for a time and then rub with aqua ammonia diluted with water. The oil and ammonia form a soap which readily washes off, bringing the rust with it.

2. Wash the stains with dilute hydrochloric acid until removed, dry and polish with tripoli.

Cleaning Nickel Plates.

Nickel plates, must, under no circumstances, be brushed with chalk, but be carefully washed with soft water and soap, with a soft brush. Any tarnish or spots can be removed by dipping the plates for a few seconds in a solution of clean and moderately strong cyanide of potassium, rinse in clean water, dip in alcohol, and dry the work in clean boxwood sawdust. When dry, remove any sawdust that may remain with a camel's hair brush. Handle the plates with tissue paper. Carefully buff the plates for a finish with soft rough buff. The buff must be free from dust. The above must be strictly adhered to, in order to have good work.

Restoring the Color of Nickel Movements.

Take 50 parts of rectified alcohol, 1 part of sulphuric acid and 1 part nitric acid. Dip the pieces for about 10 to 15 seconds in this composition, then dip them in cold water, and afterwards in rectified alcohol. Dry them with a piece of fine linen, or in sawdust. Nickel, and the greater part of those metals liable to tarnish, may be restored to their primitive color by dipping in the following bath: Dissolve in a half a glass of water, 6 or 7 grams of cyanide of potassium; plunge the pieces in this solution and withdraw them immediately. As the cyanide mixes well with water, it is sufficient to rinse them once in the latter to destroy any trace of the cyanide. After this, dip the pieces in alcohol, and dry them in boxwood dust, in order to keep them from rusting. The balance, even together with its spring, can be subjected to this operation without any danger. If the pieces to be restored are greasy, they must be cleaned with benzine before being dipped in the cyanide, because it will not touch grease. Cyanide of potassium, being a violent poison, great care has to be exercised, and the operation should be performed in a well ventilated place. The same bath can be preserved in a bottle, and serves for a long time.

Polishing Aluminium.

Mix equal parts of rum and olive oil, by shaking these liquids together in a bottle. When a burnishing stone is used, the peculiar black streaks first appearing should not cause vexation, since they do not injure the metal in the least, and may be removed with a woolen rag. The object in question may also be brightened in potash lye, in which case, however, care must be taken not to make use of too strong lye. For cleaning purposes benzole will be found best. Objects of aluminium can be electroplated without any difficulty, and a bright white luster may be imparted to them by passing them successively through a weak bath of hydrochloric acid and aquafortis. The effect obtained is quite surprising.

Cleaning Clocks.

If an American clock need no repairs, just cleaning, it is not best to take apart but proceed as follows: Procure a flat sash or varnish brush (new) 1 inch wide, and a square tin pan 8x8 inches, and say 2 inches deep. Wind up the clock and hold the movement over your tin pan, and with the sash brush referred to, apply common kerosene oil. Remove the pallet (verge) and allow the clock to run down rapidly once or twice, applying freely all the while the oil to all the pivots, pinions, etc. This will cut and remove the gum on the springs and pivots. Now to clean off the oil, throw out the oil and use 74 degrees gasoline in the same manner, rinse freely with gasoline and allow to dry for a hour or two and your job is completed and well done. After drying one or two hours you can proceed to oil the neeessary parts, but never oil the mainsprings. If a clock needs repairing always put it through this process before taking it apart, for it is then clean to handle. A clock can be cleaned fifty times by this process and never injure the lacquer. If a clock is not very dirty, gasoline (benzine will do) alone will answer without the kerosene oil. French clocks should always be carefully taken down, and put through the same process, but each piece wiped clean with chamois skin.

Bleaching Ivory.

Ivory that has become yellow, may easily be bleached in the following manner: The article is placed under a glass bell, together with a small quantity of chloride of lime and muriatic acid, whereby chlorine is developed, and exposed to sunlight. Be very careful not to breathe the vapors, as they are very poisonous. The bleaching power of the chlorine destroys the yellow pigment upon the surface, and the article will be restored to its original luster.

Cleaning Ivory Ornaments.

Ivory ornaments are quickly cleaned by brushing them with a new, not very sharp, tooth brush, to which a little soap has been given, then rinse the ornament in lukewarm water; next dry it and brush a little, and continue brushing until the luster reappears, which can be increased by pouring a little alcohol upon the brush, and applying it to the object. Should this have become a little yellow, dry it in gentle heat, and it will appear as if new.

Removing Stains from Watch Dials.

To remove black or cloudy stains from porcelain watch dials, which are generally caused by the tin boxes they are shipped in, wet a piece of tissue paper in nitric acid and wipe the dial. This will instantly remove them. After applying the acid the dial should be immediately washed thoroughly in water and then dried in boxwood sawdust.

Fine Rouge.

Dissolve sulphate of iron in hot water until no more will be taken up. Allow to settle and pour the clear liquor into a large jar, filling it only about half full. Add dilute oxalic acid, slowly stirring with a glass rod until the yellow precipitate ceases to form. Filter and dry the resulting precipitate; drive off the oxalic acid by exposure to moderate heat. The resulting ferric oxide will be very fine and pure.

French Polishing Powder.

Mix 1 part of fine rouge with 50 parts of carbonate of magnesia, moisten a rag with water or alcohol, dip it into the powder, and rub the articles thoroughly. Dry them with soft leather.

Putty Powder.

Put pure metallic tin in a glass vessel and pour in sufficient nitric acid to cover it. Great heat is evolved with considerable effervescence, so that care should be taken that the vessel is sufficient large to prevent boiling over. The fumes are poisonous. When nothing is left but a white powder, dry at a gentle heat to drive off the free acid.

Tripoli.

A gray-white or yellowish powder, which is made from the shells of microscopic organisms. It is used for polishing soft metals, first with oil and then dry.

Pulz Pomade.

Oxalic acid 1 part, oxide of iron 25 parts, rotten stone 20 parts, palm oil 60 parts, vaseline 4 parts. The oxide of iron may be Venetian red, or fine rouge, according to quality desired. All the powders must be absolutely free from grit.

Pickling Solutions.

1. Sulphuric acid 1 part, water 8 parts.

2. Muriatic acid 1 part, water 8 parts.

3. Sulphuric acid 1 part, nitric acid 1 part, water 2 parts.

4. Dilute aqua regia will remove tarnish from gold.

5. Acetic acid 1 part, water 4 parts. Plunge the article into the pickle while hot if quick action is desired, either articles or pickle may be heated.

Cold Black Pickle for Brass.

All heretofore known black and gray pickles possess the defect that they give different colors with different copper alloys, while in the case of certain alloys they altogether refuse to act. For instance, carbonate of copper, dissolved in ammonia, gives to brass a handsome, dark gray color, while it does not attack various other alloys; but it is little suitable for instruments. A dark-gray pickle, which almost indiscriminately stains all copper alloys a handsome gray, resembling in color the costly platinum, is composed by dissolving 50 grams arsenic in 250 grams hydrochloric acid, and adding to the solution 35 grams chloride of antimony and 35 grams finely pulverized hammer scales. The articles to be pickled are rinsed in a weak, warm soda solution, prior to, as well as after immersion, to be followed by continued rinsing in water. The recipe is simple, and has been repeatedly tested with uniformily good results.

Pickle for German Silver.

To twelve parts of water add one part of nitric acid, immerse the article in this, quickly remove, and place in a mixture of equal parts of sulphuric and nitric acids, rinse thoroughly in water, and dry in sawdust. In all cases of pickling it is essential that all traces of acid be removed by frequent washings in clean water.

Pickling of Metals.

Metals are pickled for the purpose of removing the oxides and producing a lustrous surface. An excellent pickle for brass consists of 10 parts of water and 1 of sulphuric acid. Dip into this pickle, wash, dry, and immediately dip into a second pickle consisting of 2 parts nitric acid and 1 of sulphuric acid and rinse thoroughly. This dissolves the zinc from the brass, and gives the metal a brilliant surface. All pickling operations with either hot or cold pickle should be carried on in the open air or in the draft of a well drawing chimney, as the vapors arising from the acids are very injurious. In order to retain the luster, a good transparent varnish should be applied.

Pickle for Gold Alloys.

Gold alloys, especially those containing copper, assume an unsightly dark brown exterior, owing to the copper oxide generated by the repeated glow-heating during work. In order to remove this, the object must be pickled, and either highly diluted sulphuric or nitric acid is used for the purpose, according to the color the article is designed to have.

If working with an alloy consisting only of gold and copper, either sulphuric or nitric acid may be used indefinitely, since gold is not attacked by either of these acids, while copper oxide is easily decomposed thereby, and after having been pickled, the article will assume the color of pure gold, because its surface is covered with a layer of the pure metal.

If the alloy is composed of pure gold and silver, however, only nitric acid can be employed, and the article is left immersed in it only for a short time; this acid dissolves a very small portion of the silver, and the article also assumes the color of pure gold.

When working with an alloy which, besides the gold, contains both copper and silver, the process of pickling may be varied in accordance with the color desired to be given to the article. If the pickling is performed in sulphuric acid, the copper alone is dissolved, the article assuming a color corresponding to a gold-silver alloy, which now constitutes the surface of the article.

If nitric acid is used, it will dissolve the silver as well as copper, and in this case a pure gold color is produced.

Pickling is done by first feebly glow-heating the article and cooling it; this operation is for the purpose of destroying any fat from the hands or other contamination adhering to the article. If it was soldered with some easily-flowing solder, this glow-heating must be omitted, but it may be cleansed from impurities by immersing it at first into very strong caustic lye, and rinsing it with water; it is then laid into the acid.

The acids are employed in a dilute state, taking 40 parts water to 1 part concentrated sulphuric or nitric acid. If more articles than one, they had best be laid beside each other in a porcelain or stoneware dish, the diluted acid is poured over them, and some article is lifted out from time to time to watch the course of proceedings, whether it has assumed a yellow color.

When to satisfaction, they are rinsed with clean water and dried. While pickling for the purpose only of causing the color peculiar to gold to appear, the process of coloring has for its object to lend the appearance of very fine gold to an article of an indifferent alloy. Various mixtures may be employed for the purpose, and we give two receipts below which are very appropriate:

Mix two parts saltpeter, 1 part table salt and 6 parts alum with 6½ parts water, and place in a porcelain dish for heating. As soon as you notice that the mixture begins to rise, add 1 part of muriatic acid, raise the whole to boiling and stir with a glass rod.

The article to be colored, and previously treated with sulphuric acid, as specified, is suspended to a hook, either of sufficiently thick platinum wire or glass; it is then introduced into the rather slow boiling bath, and moved around in it. It is to be taken out in about three minutes, and rinsed in clean water, inspecting its color at the same time. If not to satisfaction, it is returned to the bath, and this withdrawing or reintroducing is repeated until the desired color is obtained. By the latter immersions the article is left only one minute at a time in the fluid.

When sufficiently colored, the article after rinsing, will be of a high yellow and mat color; it is washed repeatedly in water to remove the last traces of the bath, and then dried between soft and heated sawdust.

In place of drying in sawdust the article may also be dipped in boiling water, leaving it in for a few seconds; the adhering water will evaporate almost instantaneously.

The second coloring method consists in pouring water over a mixture of 115 parts table salt and 230 nitric acid, so that the salt is dissolved; it is then to be heated until a dry salt residue is again present. This residue is mixed with 172 parts fuming muriatic acid and heated to boiling, for which purpose a porcelain vessel is to be used.

As soon as the pungent odor of chlorine gas begins to evolve, the article to be colored is immersed, and left for about eight minutes in the fluid for the first time; in other respects, a similar treatment, as specified above, is also used for this method; if the article colored was polished previously, a subsequent polishing is unnecessary.

On account of the vapors evolved by the coloring baths, which are very dangerous to health, the operations should be performed either under a well-drawing flue, or what is still better, in open air.

Scratch Brushing.

Articles in relief which do not admit of the use of the burnisher are brightened by the aid of the scratch brush. The shape of the brush varies according to the article to be operated upon. Hand scratch brushes are sometimes made of spun glass, with fibres of extreme fineness and elasticity, and are used for scouring only very delicate objects. They are also made of numerous wires of hardened brass and are prepared in similar form to the glass brushes, except when purchased the ends of the wires are not cut off, the operator being expected to do so before using them. The object in leaving the wires connected being to prevent them becoming damaged. Circular scratch brushes, in which the wires are arranged radially, are used for scouring articles which will admit of their use. They are attached to the spindle of a polishing lathe, and the wires consequently all receive a uniform motion in the same direction. Scratch brushes are seldom if ever used dry, the tool and the work being constantly wet with a decoction of soap-root, marshmallow, cream of tartar, alum or licorice root. With small articles the scratch brush is held as you would a pencil, and is moved over the article with a backward and forward motion. The brushes must be carefully looked after and the wires kept straight and in good order. If they become greasy they are cleansed in caustic potash, and if they become rough they are sometimes dipped into nitric acid. With circular brushes it is well to reverse them occasionally in order to change the direction of the wires. Dirty polishing leathers should be cleaned by soaking them for an hour or two in a weak solution of soda in warm water, first rubbing the leather thoroughly with soap. Rinse thoroughly and wash in soap and water. The soap in the water will keep the leather soft and pliable. Dry them in a towel and rub thoroughly and your leathers will be much better than any new ones you can buy.

Mat Brushing.

Very excellent results are obtained by running the fine wire matting brush at about 2,300 revolutions per minute, applying rain water or sour beer diluted with water at the place where the brush strikes the work; occasionally hold a piece of sand-paper to the brush. Should the points of the brush be too straight, let them strike over a piece of wire, but do not hook them too much, as this would prevent matting. Always preserve the brush in a good condition; should the wires become entangled or twist into knots, separate or cut them out. After the work is matted, take a soft hair brush and brush it in soap water, then rinse it in warm water charged with a small quantity of spirits of ammonia and caustic potash; immerse it in pure alcohol for a short time, and finally dry it in sawdust.

Burnishers.

The surface to be burnished must be free from scratches which the burnisher would not remove, but render more distinct by contrast, and the burnisher must be kept highly polished, for the surface burnished can never be smoother than the burnisher. Burnishing polished pivots with the glossy burnisher preserves them from wearing, Very little, if any, of the metal is removed by burnishing in the ordinary way, although watchmakers sometimes use what are called cutting burnishers to form pivots. The cross section of these burnishers matches the outline of the pivot it is desired to form, and they are roughened by rubbing on a lead block charged with coarse emery. The pivot is finished with a smooth burnisher of the same form as the cutting one. Silversmiths use burnishers of agate.

Burnishing Powder.

A good burnishing powder is prepared from ½ pound white chalk, 2 ounces pipe clay, 2 ounces white lead, ½ ounce magnesia carbonate, and colored with the same quantity of jewelers' rouge. It is said to be unrivaled for cleansing silver.

Cleaning Rags.

These rags, which are excellent for polishing metal surfaces, are prepared in the following manner: Dip flannel rags into solution of 20 parts dextrine and 30 parts oxalic acid in 20 parts logwood decoction, wring them gently, and sift over them a mixture of finely pulverized tripoli and pumice stone. The moist rags are piled above each other, placing a layer of the powder between each two. They are then pressed, taken apart, and dried.

Cleaning Powder for Show Windows.

A good cleaning powder which leaves no dirt in the joints, etc., is prepared by moistening calcined magnesia with pure benzine so that a mass is formed sufficiently moist to let a drop appear when pressed. The mixture is to be preserved in glass bottles with ground stoppers, in order to retain the easily volatile benzine. A little of the mixture, when to be used, is placed upon a lump of cotton and applied to the glass plate.

Cleaning Soiled Chamois Leather.

Many workshops contain a dirty wash leather, which is thrown aside and wasted for want of knowing how to cleanse it. Make a solution of weak soda and warm water, rub plenty of soft soap into the leather, and allow it to remain in soak for two hours, then rub it well until quite clean. Afterward rinse it well in a weak solution composed of warm water, soda and yellow soap. It must not be rinsed in water only, for then it will be so hard, when dry, as to be unfit for use. It is the small quantity of soap left in the leather that allows it to separate and become soft. After rinsing, wring it well in a rough towel, and dry quickly, then pull it about, and brush it well, and it will become softer and better than most new leathers. In using a rough leather to touch up highly polished surfaces, it is frequently observed to scratch the work; this is caused by particles of dust, and even hard rouge, that are left in the leather, and if removed by a clean brush containing rouge, it will then give the brightest and best finish, which all good workmen like to see on their work.

Cleaning Brushes.

The best method of cleansing watchmakers' and jewelers' brushed, is to wash them out in a strong soda water. When the backs are wood, you must favor that part as much as possible, for being glued the water may injure them.

Composition Files.

These files, which are frequently used by watchmakers and other metal workers, for grinding and polishing, and the color of which resembles silver, are composed of 8 parts copper, 2 parts tin, 1 part zinc, 1 part lead. They are cast in forms and treated upon the grindstone; the metal is very hard, and therefore worked with difficulty with the file.

To Prepare Chalk.

Pulverize the chalk thoroughly and then mix it with clean rain water, in proportions of two pounds to the gallon. Stir well, and then let it stand about two minutes. In this time the gritty matter will have settled to the bottom. Slowly pour the water into another vessel, so as not to stir up the sediment. Let stand until entirely settled, and then pour off as before. The settlings in the second vessel will be prepared chalk, ready for use as soon as dried. Spanish whiting, treated in the same way, makes a very good cleaning or polishing powder. Some watchmakers add a little crocus, and we think it an improvement; it gives the powder a nice color, at least, and therefore adds to its importance in the eyes of the uninitiated.

Diamantine.

Diamantine consists of crystalized boron, the basis of borax. By melting 100 parts boric acid and 80 parts aluminium, crystals are obtained, the so-called bort, which even attacks diamond. Diamantine bought in commerce is less hard.

Vienna Lime.

A pure, anhydrous lime, obtained from Vienna. It is extensively used for final polishing purposes, particularly in watch factories. It differs from most polishing substances in that the effect is not produced by simple abrasion, for unless this lime is used while it is slacking, the result will be unsatisfactory. It is kept in tightly corked bottles and no more than is wanted taken out at one time. Take a small lump from the bottle, crush, mix to a paste and rub on the article with a boxwood slip, using quick strokes.

SOLDERS AND SOLDERING.

Soldering.

Soldering is the act of joining two metallic surfaces by means of a more fusible metal or metallic cement. Solders are commonly divided into two groups known as hard solders and soft solders; the former fuse only at a red heat, while the latter fuse at low degrees of heat. In hard soldering it is frequently necessary to bind the parts to be soldered together with what is known as binding wire, which is made of soft iron, repair clamps or soldering forceps. The blowpipe is used most extensively for soldering, although small soldering irons are used on the larger kinds of work. It is of the utmost importance that the meeting edges of all articles to be soldered be scraped or chemically cleaned. While soldering, articles are usually placed upon a piece of charcoal, though asbestos or pumice stone is better for the purpose. Charcoal emits gases from the coal while under the blowpipe which enter into the alloy of gold or silver and render it brittle. To prove this, reduce a small piece of 10k gold to a liquid form on a piece of charcoal, and treat a piece simialrly on a piece of asbestos or pumice stone, and after allowing each to cool, subject both to a heavy pressure and note the difference in their malleability and ductility.

Soldering Forceps or Repair Clamps.

By the use of these devices any article to be repaired can be adjusted in any desired position in a much shorter time, and with more accuracy than by the ordinary process of binding with wire to a piece of charcoal. They are so constructed that any two pieces can be as readily brought together as can be done with the fingers, no matter at what angle or position you may desire them. Each part works independent of the other, and the whole is held securely in place by means of nuts, as shown, and both hands being free, charcoal can be held behind the article,

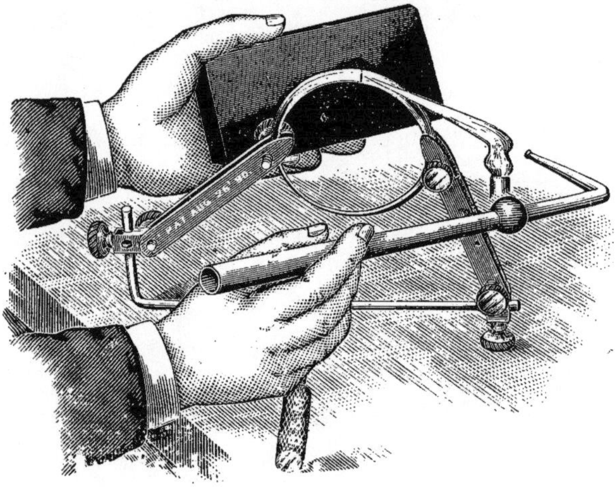

thereby concentrating the heat, the same as when held directly upon it. In soft soldering it can be used to great advantage.

The forceps revolve in parts, which are fastened to arms, by means of a hinge joint. The arms run through the collars, so that they can be lengthened or shortened, and the forceps raised or lowered as desired. The collars turn independently of each other on base, and being split the whole is held firmly in position by nuts.

Gold Solders.

Gold solders should approach the articles to be soldered in both color and fusibility as nearly as possible. The following gold solders are in general use:

	Parts Gold	Parts Silver.	Parts Copper.	Parts Zinc.
Hard solder for 750 fine	9 0	2.0	1.0	
Soft solder for 750 fine	12.0	7.0	3.0	
Solder for 583 fine	3 0	2.0	1.0	
Solder for less than 583 fine	2.0	2.0		
Readily fusible solder	11 94	54.74	28.17	5.01
Solder for yellow gold	10.0	5.0		1.0

To Solder Enameled Jewelry.

If the enamel is good, hard enamel, heating to melt the hard solder will not affect it much. The trouble usually is that the article is thrown when nearly red hot into the pickle. Enameled articles should be allowed to cool and then boiled in a copper dish with sulphuric acid, 1 part; water, 5 parts.

Enamel Solders.

1. Copper 25 parts, silver 7 parts, gold 68 parts.

2. Silver 18 parts, gold 74 parts.

The above solders are very refractory, and are used for work that is to be subsequently enameled.

To Preserve the Color of Gold.

To preserve the color of gold while soldering is very important. One ounce of yellow ochre mixed with borax water to the consistency of cream, and applied to the article before heating will overcome the difficulty. After heating boil out as usual.

To Prevent Discoloring of Silver.

A paste of whiting and water, dried on the bright parts of silver, will prevent oxidation while soldering. Care should be taken to keep the paste from touching the parts that have been cleaned for the solder.

Silver Solders.

The following hard silver solders have been thoroughly tested:

	Parts Fine Silver.	Parts Copper.	Parts Brass	Parts Zinc.
First	4		3	
Second	2		1	
Third	19	1	10	5
Fourth	57	28.6		14.3

Yellow Solder for Brass.

1. Copper 1 part, zinc 1 part.
2. Copper 32 parts, zinc 29 parts, tin 1 part.
3. Zinc 3 parts, copper 2 parts.

To Solder a Stay Spring.

Stay or lifting springs are often broken, and the watchmaker has frequently none of the right size nor the time to make a new one. In such a predicament he can mend the old one and have it just as good as new, by placing the broken parts together and binding them firmly to a piece of coal, then soldering them with 18-karat gold. It requires a strong heat and plenty of borax; then finish off, nicely harden and temper in the usual manner.

To Solder Broken Broaches.

Steel broaches and other tools are soldered by cleaning well the parts broken, then dipping them into a solution of sulphate of copper, and soldering them with ordinary soft solder. The joint is a good one and will stand ordinary hard wear.

Solder for Aluminium.

The following alloys are recommended for the purpose:

1. Melt 20 parts of aluminium in a suitable crucible, and when in fusion add 80 parts zinc. When the mixture is melted, cover the surface with some tallow, and maintain in quiet fusion for some time, stirring occasionally with an iron rod; then pour into moulds.

2. Take 15 parts of aluminum and 85 parts zinc, or 12 parts of the former and 88 parts of the latter, or 8 parts of the former and 92 parts of the latter; prepare all of them as specified for No. 1. The flux recommended consists of three parts balsam copaiba, one of Venetian turpentine, and a few drops of lemon juice. The soldering iron is dipped into this mixture.

To Solder German Silver.

Dissolve granulated zinc in muriatic acid in an earthen vessel. Cleanse the parts to be soldered and apply the acid. Next put a piece of pewter solder on the joint and apply the blowpipe to it. Melt German silver one part, and zinc in thin sheets four parts; then powder it for solder.

Soldering Stone Set Rings.

There are various ways for doing this, but the following will be found to be as good as any: Take tissue paper and tear it into strips about three inches wide, twist them into ropes, and then make them very wet and wrap the stone with them, passing around the stone and through the ring until the center of the ring is a little more than half full of paper, always winding very close, and then fasten upon charcoal, allowing the stone to project over the edge of the charcoal, and solder very quickly. The paper will prevent oxidation upon the part of the ring it covers, as well as protect the stone.

Soft-Soldering Articles.

Moisten the parts to be united with soldering fluid; then, having joined them together, lay a small piece of solder upon the joint, and hold over your lamp, or direct the blaze upon it with your blowpipe until fusion is apparent. Withdraw them from the blaze immediately, since too much heat will render the solder brittle and unsatisfactory. When the parts to be joined can be made to spring or press against each other, it is best to place a thin piece of solder between them before exposing to the lamp. Where two smooth surfaces are to be soldered one upon the other, you may make an excellent job by moistening them with the fluid, and then, having placed a sheet of tin foil between them holding them pressed firmly together over your lamp till the foil melts. If the surface is fitted nicely, a joint may be made in this way so close as to be almost imperceptible. The bright looking lead, which comes as a lining of tea boxes, is better than tin foil.

To Dissolve Soft Solder.

Nitric acid may be used safely for gold not lower than 12 karat and is very effective. The following is suitable for all grades of gold and silver: Green copperas 2 ounces, saltpeter 10 ounces, reduced to a powder and boiled in 10 ounces of water. It will become crystalized on cooling. Dissolve these crystals by the addition of 8 parts of spirits of salts to each part of crystals, using an earthenware vessel. Add 4 parts of boiling water, keep the mixture hot, and immerse the article to be operated upon, and the solder will be entirely removed without injuring the work.

Soft Solder.

The soft solder most frequently used consists of 2 parts of tin and 1 of lead. The following table gives the composition of various soft solders with the respective melting points:

Number.	Parts Tin	Parts Lead.	Melts at Degrees F.	Number.	Parts Tin.	Parts Lead.	Melts at Deg. F.
1	1	25	558°	7	1½	1	334°
2	1	10	541	8	2	1	340
3	1	5	511	9	3	1	356
4	1	3	482	10	4	1	365
5	1	2	441	11	5	1	378
6	1	1	37	12	6	1	380

Hard Solders.

Under this name very different alloys are used, depending upon the metals to be united. The following table shows the composition of various hard solders which have stood a practical test for various purposes:

	Part Brass.	Parts Zinc.	Parts Tin.
Refractory	4.00	1.00	
Readily Fusible	5 00	4 00	
Half White	12.00	5.00	1.00
White	40 00	2.00	8 00
Very Ductile	78.00	17.25	

To Make Wire Solder.

Melt the solder and pour into a vessel having a very small opening, over a pail of water. The molten metals will run through the hole into the water, solidfying as they pass. The vessel should be hot enough to avoid chilling solder until after it leaves the hole.

Soldering Fluxes.

For hard solder use borax rubbed to a paste with water on a slate. For soft soldering dissolve a small piece of zinc in pure hydrochloric acid until effervescence ceases. Take out the undissolved zinc after twenty-four hours, filter the solution, add ⅓ its volume of spirits of sàl ammoniac and dilute with rain water. This fluid is non-corrosive.

Jewelers' Soldering Fluid.

Add to alcohol as much chloride of zinc as it will dissolve. Apply with a stick or medicine dropper.

Non-Corrosive Soldering Fluid.

A non-corrosive soldering fluid is prepared in the following manner: Small pieces of zinc are immersed into muriatic acid to saturation, which can be known by the cessation of the ebullition; the zinc, also, being added after this point, remains undissolved; add about one-third the volume of spirits of ammonia, and dilute with a like quantity of rain water. If the acid is gently heated at the time of adding the zinc, the dissolving will progress much more rapidly. This fluid causes no rust on iron or steel, and is even excellent for tinning.

Etching.

Etching is accomplished by eating away certain portions of the surface of the metal by acids, while the balance is protected by a coating of wax or other substance impervious to acids. It may be done in relief or intaglio, according as the design is required; in relief, by etching the back ground and leaving the higher portions bright; or in intaglio, by covering the plate with a coating of wax or special ink and drawing or scratching the design upon the metal by means of variously shaped points called etching needles. In the latter case, the points are handled much like a lead pencil, care being taken to always cut through the wax. The work may be left the original color of the metal, with a mat surface, or it may be colored to throw the design in relief by using any of the well-known oxidizing solutions before removing the wax.

For single pieces slightly warm the article to be etched and draw the design on the metal with a steel pen and asphaltum or a pointed stick of wax, then plunge in the acid or lay the acid on the design with a camel hair pencil, as is most convenient, repeating the applications until the work is etched sufficiently deep, then wash in water and remove the wax with turpentine, benzine or alcohol.

Where large quantities of work are to be done, such as spoons, trade marks on steel goods, etc., a rubber

or composition stamp is made, having the design made so that the parts to be bitten by the acid shall be depressed in the stamp as shown in the illustration.

These stamps are easily made as follows: Sketch the design or word upon a piece of paper and when satisfactory draw it upon the inside of the bowl of the spoon with plaster of paris. Mix the plaster to a consistency of cream, and apply it by means of a small camel's hair brush or by means of a very small syringe. The plaster should be laid on rather heavily, in order that the depressions in the stamp shall be sufficiently deep. After drawing is completed and thoroughly hardened, oil it, and the bowl of the spoon with olive oil, applying with a small brush. Procure a small piece of composition such as printer's rollers are made from, (a mixture of glycerine, glue and molasses), heat it over a water bath until quite liquid and then pour into the bowl of the spoon. Put away until thoroughly cool. This composition pad can then be removed, glued to a wooden handle, such as are used for rubber stamps and you are ready to do the printing on any number of spoons. The receipt for stamp ink will be found elsewhere. If all the operations have been performed properly you will have a stamp similar to the illustration, in which the name "Aurora" appears in white, the letters being countersunk in the face of the stamp. With a small roller or pad distribute your stamp ink thoroughly upon the surface of a small piece of glass, apply your stamp to the glass, thoroughly inking it by rolling back and forth and then apply to the bowl of the spoon. The result will be that the entire bowl of the spoon is covered with the ink except the word "Aurora," which is left white where the silver is exposed. You are now ready to proceed with your etching. If your stamp has good depth, the composition is of the proper consistency and the ink well distributed, the lettering will need no touching up by hand, if it should, however, it may easily be done by means of a fine pointed brush and asphaltum. This stamp is used with a suitable ink or wax to transfer the design to the goods and the work proceeded with as

before. In the first illustration shown, the words "Bridgeport, Conn.," are bright, and in relief, the letters being the only part touching the wax; while in the second, the word "Aurora," was the only part not touched by the wax, and consequently it is etched in intaglio and the bowl is left bright. Almost any of the dilute acids may be used that will attack the metal to be acted upon, but a number of fluids are given here.

Grounds for Etching.

1. White wax, 30 parts; gum mastic, 30 parts; asphaltum, 15 parts.

2. White wax, 30 parts: gum mastic, 15 parts; asphaltum, 15 parts.

3. White wax, 60 parts; gum mastic, 30 parts; asphaltum, 60 parts.

4. White wax, 3 parts; block pitch, 1 part; asphaltum, 4 parts; resin, 1 part.

5. Soft linseed oil, 4 oz.; white wax and gum benzoin, each ½ oz.; boil to two-thirds.

Etching Fluids for Brass.

1. Dissolve 6 parts chlorate of potash, 100 parts water, add 160 parts water to 16 of fuming nitric acid; mix the two solutions.

2. One part sulphuric acid, 8 parts water.

3. One part nitric acid, 8 parts water.

4. Nitric or sulphuric acid 1 part, saturated solution of bichromate of potash 2 parts, water 5 parts.

Etching on Copper.

1. Fuming hydrochloric acid, 10 parts; water 70 parts; add a boiling solution of potassium chlorate and dilute.

2. Acetic acid 8 ounces; ammonium chloride, 60 gram; sodium chloride, 60 gram; pure verdigris, 40 gram; powder the solids and boil in the acid until dissolved.

3. Nitrous acid, 1 ounce; silver acetate, 3 drams; nitric ether, 8 ounces.

Etching Fluids for Silver.

Use any of the fluids for brass and copper, any of the oxidizing solutions given elsewhere, or any of the dilute acids that will attack silver. Care must be taken in stopping out the silver not to be etched, in order that the acids may not creep under and destroy the sharpness and beauty of the work.

Etching Fluids for Ivory.

1. Use dilute sulphuric and hydrochloric acids, mixed.

2. Cover the ivory to be etched with a thin coating of bees wax, then trace the figure you desire to present through the wax. Pour over it a strong solution of nitrate of silver. Let remain a sufficient length of time, then remove it, with the wax, by washing in warm water. The design will be left in dark lines on the ivory.

Etched Signs for Jewelers.

Any jeweler can make extremely elegant signs by taking sheet brass or copper, painting the parts not to be etched with asphaltum or any of the other grounds and etching as deeply as may be required. If a number of signs are to be made, paint the backs of the sheets also and plunge them in a wooden trough filled with the etching solution. A very effective design is made by leaving the letters and borders bright, etching quite deeply and then finishing with the green bronzing solution given elsewhere. Remove the asphaltum by wiping with a rag moistened with turpentine, clean thoroughly and lacquer. Or the letters only may be etched, and afterwards filled in with black or colored pigments, finishing with lacquer. An ingenious and tasty jeweler can thus secure an infinite variety of extremely elegant and serviceable signs at a very small expense, beside utilizing his leisure moments in a very interesting and fascinating way.

Engravers' Border Wax.

Beeswax, 1 part; pitch, 2 parts, tallow, 1 part, mix.

Stamp Ink.

Melt ¼ lb. resin, add 1 teaspoonful of lard oil, stir in a tablespoonful of lamp black and throughly mix; thin with turpentine to make it of the consistency of printer's ink when cold.

Wax for Steel.

Melt equal parts burgundy pitch, asphaltum and beeswax, stirring constantly until thoroughly incorporated, apply with a dabber or ball of cotton covered with silk, having first warmed the article so that the stick of the wax will readily melt by touching. Rub on the wax in the stick and spread evenly with the dabber.

HARDENING, ANNEALING AND TEMPERING.

Hardening.

Gold, silver, copper and brass are hardened by heating and allowing to cool slowly; also by beating, burnishing, etc. Steel is hardened by beating, also by heating to a cherry red and then cooling quickly, by plunging in water or some one of the numerous hardening mixtures.

The recipes here given are from various sources, and the reader must adopt the one which he finds on trial, is the best adapted to his wants.

In all cases the object should be heated to a red heat before plunging. If an object to be hardened is long and slender, it should invariably be inserted in the hardening compound endwise, otherwise it will come out warped and distorted. The same rule applies to thin or flat objects. A preparation is used in hardening, consisting of one teaspoonful of wheat flour, two of salt and four of water. The steel to be hardened is to be heated sufficiently, dipped into the mixture to be coated therewith, then raised to a red glow, and dropped into cold soft water. Another method is to raise the object to the required heat and then drop it into a mixture of ten parts mutton suet, two parts sal-ammoniac, five parts resin and thirty-five parts olive oil. Oil, tallow, beeswax, and resin are also employed for hardening. If an intense brittle hardness is desirable drop the object into mercury or nitric acid. In heating very small or thin objects, they should be placed between two thin pieces of charcoal and the whole brought to the required heat. In this way you avoid uneven heating and hence it will be uniformly tempered. When it is desirable to harden an article without discoloring its surface, it should be placed in a metal tube or bowl of a clay pipe, and surrounded with charcoal that has been previously heated to expel all moisture, and when raised to the proper heat the whole should be immersed in the hardening liquid.

To Harden Steel in Petroleum.

According to B. Morgossy, the articles to be hardened are first heated in a charcoal fire, and, after thoroughly rubbing with ordinary washing soap, heated to a cherry red. In this condition they are plunged into petroleum; ignition of the petroleum need not be feared if no flame is near at hand. Articles hardened by this method show no cracks, do not wrap if plunged endwise, and after hardening remain nearly white, so they can be blued without further preparation.

Hardening Liquids.

If water is used for hardening, 32° F. will be found about right for the sized articles hardened by watchmakers and if the article is very small, ice may be added to the water. A solution composed of one quart of water, 1¼ lbs. of sal-ammoniac, 10 oz. of refined borax, 1¼ ozs. of red wine, is used extensively for fine cutlery. A mixture of 1 lb. of resin, 3 ozs. of lard, ½ lb. train oil and ½ oz. of assafoetida is said to be excellent for fine steel work.

Combined Hardening and Tempering.

M. Caron, with a view to combining the two operations of hardening and tempering, suggested that the temperature of the water used for hardening, be heated to a pre-determined degree. Thus the requisite temper may be given to gun-lock springs by heating the water in which they are hardened to 55° C., or 130° F.

Annealing.

Gold, silver, copper and brass are annealed by heating them to a red heat and then plunging in water. Iron and steel are annealed by heating to a red heat and allowing to cool slowly, either in the open air or in various mixtures, some of which are given herewith. There are nearly as many methods of annealing as there are workmen. The commonest methods are as follows: Heat to a dull red, bury in warm iron filings or ashes, and allowing the article to cool very gradually. Another method is to heat the piece as slowly as possible, and when at a low red heat put it between two pieces of dry board and screw them up tightly in a vice. The steel burns its way into the wood, and on coming together around it they form a practically air-tight charcoal bed. Brannt gives the following method which he says will make steel so soft that it can be worked like copper: Pulverize beef bones, mix them with equal parts of loam and calves' hair and stir the mixture into a thick paste with water. Apply a coat of this to the steel and place it in a crucible, cover this with another, fasten the two together with wire and close the joint hermetically with clay. Then put the crucible in the fire and heat slowly. When taken from the fire let it cool by placing it in ashes. On opening the crucible the steel will be found so soft that it can be engraved like copper.

To Anneal Small Steel Pieces.

Place the articles from which you desire to draw the temper into a common iron clock key. Fill around it with brass or iron filings, and then plug up the open end with a steel, iron or brass plug, made to fit closely. Take the handle of the key with your plyers and hold its pipe into the blaze of a lamp till red hot, then let it cool gradually. When sufficiently cold to handle, remove the plug, and you will find the article with its temper fully drawn, but in all other respects just as it was before. The reason for having the article thus plugged up while passing it through the heating and cooling process is, that springing always results from the action of changeable currents of atmosphere. The temper may be drawn from cylinders, staffs, pinions, or any other delicate pieces by this mode with perfect safety.

To Soften a Spring.

A spring may be laid flat and its temper drawn between two plates fastened together by a screw through the center, and placed upon the annealing plate. A small piece of whitened steel is laid upon it, to enable the operator to judge of the degree of heat. Before opening, let it cool. When drawing the temper lay the coils farther apart.

To Soften Steel.

Heat it brown-red, and plunge it in soft water; river water is best. Do not heat over red-brown, however, else it becomes hard when plunged. But if you plunge it as soon as it turns red, the steel will be soft enough to cut with ease.

To Anneal Hardened Steel.

It may sometimes happen, that hardened steel parts require a few finishing touches, which can not be done because they are too hard and their polish would be ruined by annealing them, because it turns blue, and the piece then requires renewed polish, which consumes a great deal of time. The most practical way then is to cover the steel part with the oily dirt from the oil-stone, after which it can be annealed with impunity, that is, the flame is, with the blow-pipe, directed to the point required. The article is afterward cleansed in benzine.

To Anneal a Staff or Pinion.

It sometimes becomes necessary to anneal a staff or pinion, in which you wish to insert a pivot, without removing it from the wheel. To do this place the whole part or end of the staff or pinion in a pin-vise or slide-tongs, which, of course, is cold; now pierce the top of a brass thimble, so that the end to be drilled will go in snug; then, with a blowpipe and small spirit lamp throw as much heat as you wish on the article to be drilled, by blowing directly into the thimble, without in the least heating the wheel.

Tempering.

Before tempering, the surface of the object must be thoroughly cleaned and freed from grease by the application of oilstone dust, emery, or some like scouring agent. The object should not be handled with the fingers after cleaning, or it will be difficult to obtain the requisite tint.

After letting an object down to the required color it should be allowed to cool gradually, and no artificial means employed to hasten the cooling. A piece of steel may be let down to the same color several times without in any way injuring it or altering its properties. Tempering of small articles is performed satisfactorily by means of the bluing pan. Small articles are also tempered by placing them in a vessel, say a large spoon, covering them with oil and heating them to the requisite degree. This is a favored method of tempering balance staffs and similiar articles.

The following table by Stodart will be valuable to the student:

1	430° F.	Very Pale Straw Yellow	220° C.
2	450° F.	A Shade Darker Yellow	235° C.
3	470° F.	Darker Straw Yellow	245° C.
4	490° F.	Still Darker Straw Yellow	255° C.
5	500° F.	Brown Yellow	260° C.
6	520° F.	Yellow tinged with Purple	270° C.
7	530° F.	Light Purple	275° C.
8	550° F.	Dark Purple	290° C.
9	570° F.	Dark Blue	300° C.
10	590° F.	Paler Blue	310° C.
11	610° F.	Still Paler Blue	320° C.
12	630° F.	Light Bluish Green	335° C.
13	980° F.	Incipient Red	525° C.
14	1290° F.	Dull Red	700° C.
15	1470° F.	Incipient Cherry Red	800° C.
16	1650° F.	Cherry Red	900° C.
17	1830° F.	Clear Cherry Red	1000° C.
18	2010° F.	Deep Orange	1100° C.
19	2190° F.	Clear Orange	1200° C.
20	2370° F.	White	1300° C.
21	2550° F.	Bright White	1400° C.

The temper is usually judged by the color of the smoke; Saunier gives the following rule: When smoke is first seen to rise the temper is dark yellow, (or No. 2). Smoke more abundant and darker, (No. 5). Black smoke still thicker, (No. 7). Oil takes fire when lighted paper is presented to it at No. 9. After this the oil takes fire of itself and continues to burn. If the whole of the oil is allowed to burn away No. 12 is reached.

Bluing Pan.

A pan used for bluing screws and other small articles. It is sometimes very desirable to match the color of screw heads in a watch. By making the following described simple little tool you can very readily color your screws straw, purple or blue as the case may require, to match the other screw in the watch. Select a very large main-spring barrel, drill a hole in the side of the barrel the size of an ordinary pendulum rod for an American clock, cut a thread in this hole and also on the piece of wire and screw it firmly into the mainspring barrel, cutting off about four or five inches long, to which attach a neat piece of wood to serve as a handle. Now take out the head, and fill the barrel full of fine marble dust or brass or iron fillings and replace the head in the barrel, after which drill any number and size of holes in the barrel you wish, to accommodate all sizes of watch screws, and the tool is ready for use. Bluing pans similar to the one shown, can be purchased from material dealers and are similar to the one described. After fitting the screw to the proper place in the watch, harden and temper in the usual manner. Polish out all scratches or other marks and selecting a hole in the tool to fit the screw loosely, press it down level with the face of the barrel and hold the tool over a small alcohol lamp flame until the color desired appears. Heat up slowly and the effect will be much better than if it is done rapidly. First blue the screws without any special regard as to uniformity of color. Should they prove to be imperfect, take a piece of clean pith and whiten the surface with rouge, without letting it be too dry. Pieces when thus prepared, if cleaned and blued with care, will assume a very uniform tint.

Soft screws are sometimes very difficult to blue evenly, but this difficulty may be overcome by finishing them with a slightly soapy burnisher. Pieces that are not flat will rarely assume an even color when placed in a flat pan. To overcome this difficulty, sprinkle the bottom of the pan with fine brass filings or marble dust and press the article into it. The bluing pan or shovel should be thoroughly warmed before the articles are placed in it, in order that any moisture present may be dispersed. The pan will also be found useful for tempering small steel articles by boiling them in oil.

To Temper Small Steel Articles.

The tempering of small drills, for drilling holes in arbors, staffs, etc., which we find are very hard and difficult to perforate, may be effected in the following manner: After having filed the drill to its proper size (being careful not to flatten the cutting face), you then warm it moderately, not allowing it to become red, and run it into borax. The drill is thus coated over with a crust of borax and secluded from the air. Now it may be hardened by heating it only cherry red; after this it is inserted into a piece of borax, or what is better still, plunged it into mercury; taking care not to breathe the mercury fumes. Drills prepared in this way, without being brittle, will become exceedingly hard and the watchmaker will be enabled to drill articles which could not otherwise be perforated with a drill. Do not use broken broaches to make your drills as the steel in them is often burned, rendering the metal unfit for use in small tools. In order to make the quality of your drill a certainty, always take a new piece of round steel for the purpose.

Tempering Magnets.

M. Ducoetet uses the following process for tempering and magnetizing steel to be used as magnets. Two soft iron pole pieces are placed in the bottom of a water tight vessel and are connected with the poles of a powerful electro-magnet. The vessel is partially filled with water, and oil is poured into the vessel, which floats upon the surface of the water. The red hot bar is then passed through the liquids and comes in contact with the magnets. This softens the steel without depriving it of its power of being magnetized.

To Temper Drills.

Select none but the finest and best steel for your drills. In making them, never heat higher than a cherry red, and always hammer till nearly cold. Do all your hammering in one way, for if, after you have flattened out your piece, you attempt to hammer it back to a square or round, you will ruin it. When your drill is in proper shape, heat it to a cherry red and thrust it into a piece of resin or into mercury. Some use a solution of cyanuret of potassia and rain water for tempering their drills, but the resin or mercury will give better results.

To Bend Tempered Steel.

It very frequently happens to the repairer that he desires to bend a spring, but fearing the risk of breaking it, abandons the idea. In such a case the following hint may be useful: Suppose it is desirable to bend a side click spring of a Swiss bridge watch, which, by the way, is generally made of poor steel. Lay hold of the end in which the screw goes with a pair of brass-nosed sliding tongs, holding it in the left hand; then press a piece of brass against the click, bending it in the direction desired, and, at the same time, holding it over the flame of a spirit lamp until the center or spring part becomes a straw or dark red color. This will have the desired effect. The fact that spring-tempered steel is brought to a dark red blue twenty times over, will not reduce it below its former temper; on the contrary, it will tend to equalize and improve the temper and render it less liable to break.

Again, suppose a cylinder pivot, or any pivot, on any of the escapement parts are bent and you wish to straighten it by this process: Take a small brass bushing, fit it to the pivot and hold over the flame of the lamp, bending it at the same time in the desired direction.

RECOVERY AND REFINING OF WASTE.

Refining Sweepings.

The sweepings of the workshop contain quite a quantity of gold and silver. To 8 ozs. of the dirt, which has been washed and burnt, add salt, 4 ozs.; pearl ash, 4 ozs.; red tartar, 1 oz.; saltpeter, ½ oz.; mix thoroughly in a mortar, melt in a crucible and dissolve out the precious metals in a button.

To Recover Gold from Mixed Metals.

Gold should dissolve in a mixture of 1 part nitric and 3 parts of hydrochloric acid. Both should be chemically pure. The residue left after digestion is silver. Withdraw the clear solution and add to it a little sulphuric acid; if any precipitate falls, it is probably lead. Then the gold may be precipitated by any of the reagents and recovered by fusing in the ordinary manner. Afterwards dissolve the silver by treatment with nitric acid and precipitate it, wash dry and fuse. The residues may be thrown away after the recovery of the gold and silver.

To Separate Gold from Silver.

The alloy is to be melted and poured from a height into a vessel of cold water, to which a rotary motion is imparted. By this means the alloy is reduced to a finely granular condition. The metallic substance is then treated with nitric acid, and gently heated. Nitrate of silver is produced, which can be reduced by any of the known methods; while metallic gold remains as a black mud, which must be washed and smelted.

Melting Gold.

In melting gold use none other than a charcoal fire, and during the process sprinkle saltpeter and potash into the crucible occasionally. Do not attempt to melt with stone coal, as it renders the metal brittle and otherwise imperfect.

In melting brass alloyed gold, urge the fire to a great heat and stir the metal with the long stem of a tobacco pipe, to prevent honeycombing. If steel or iron filings get into gold while melting, throw in a piece of sandiver the size of a common nut; it will attract the iron or steel from the gold into the flux, or, sublimate of mercury will destroy the iron or steel. To cause gold to roll well, melt with a good heat, add a tablespoonful of sal ammoniac and charcoal, equal quantities, both pulverized, stir up well, put on the cover for two minutes, and pour.

Fusing Gold Dust.

Use such a crucible as is generally used for melting brass; heat very hot, then add your gold dust mixed with powdered borax. After a while a scum or slag will rise to the surface, which may be thickened by the addition of a little lime or bone ash. If the dust contains any of the more oxidizable metals, add a little saltpeter, skim off the slag or scum very carefully; when melted grasp the crucible with strong iron tongs, and pour immediately into cast iron molds, slightly greased. The slag and crucible may be pulverized, and the auriferous matter recovered by cupellating by means of lead.

To Refine Gold.

If you desire to refine gold from the baser metals, swedge or roll it out very thin, then cut into narrow strips and curl up so as to prevent its lying flatly. , Drop the piece thus prepared into a vessel containing good nitric acid, in the proportion of acid, 2 ounces, and pure rain water ½ ounce. Suffer to remain until thoroughly dissolved which will be the case in from one-half to one hour. Then pour off the liquid carefully, and you will find the gold in the form of yellow powder, lying at the bottom of the vessel. Wash it with pure water until it ceases to have an acid taste, after which you may melt and cast into any form you choose. Gold treated in this manner may be relied on as perfectly pure.

Recovering Gold from Coloring Bath.

Dissolve a handful of sulphate of iron in boiling water, and add it to your "color" water; it precipitates the small particles of gold. Now draw off the water, being very careful not to disturb the auriferous sediment at the bottom. You will now proceed to wash the sediment from all trace of acid with plenty of boiling water; it will require three or four separate washings, with sufficient time between each to allow the water to cool and the sediment to settle, before passing off the water. Then dry in an iron vessel by the fire and finally fuse in a covered crucible with a flux.

Gold and Silver from Textiles.

Cut into pieces the gold or silver lace, tie it tightly, and boil in soap lye till the size appears diminished, take the cloth out of the liquid, and after repeated rinsings of cold water, beat it with a mallet to draw out the alkali. Open the linen, and the pure metal will be found in all its beauty.

Removing Tin from Gold.

Much depends upon the treatment the tin received in soldering. If it received too much heat, it has penetrated into the gold, and can never be expelled again. This is known when scratching the tin, if it is glass hard it has become incorporated. If, however, it is still soft, scrape it off as closely as possible, and lay the article in a dilute mixture of sulphuric acid and water, and leave it immersed for a few hours. Have a care to have the fluid only strong enough to dissolve the tin, but not to attack the gold. When, after taking out, it should still show black spots, which is a sign that more tin is present scrape and immerse again.

Removing Gold.

Gold is taken from the surface of silver by spreading over it a paste made of pulverized sal ammoniac with aquafortis, and heating it till the matter smokes and is nearly dry, when the gold may be separated by rubbing it with a scratch-brush.

Separating Silver.

The silver holding alloy or metals are dissolved in the least possible quantity of crude nitric acid. The solution is mixed with a strong excess of ammonia and filtered into a high cylinder, provided with a stopper. A bright strip of copper, long enough to project beyond the liquid, is next introduced, which quickly causes separation of pure metallic silver. The reduction is completed in a short time, and the reduced silver washed first with some ammoniacal solution and then with distilled water. The more ammoniacal and concentrated the solution, the more rapid the reduction. The strip of copper should not be too thin, as it is considerably attacked, and any little particles which might separate from a thin sheet would contaminate the silver. The operation is so simple that it seems preferable to all others for such operations as the preparation of nitrate of silver from old coins, etc. Any accompanying gold remains behind during the treatment of the metal or alloy with nitrate acid, chloride of silver, produced by the impurities in the nitric acid is taken up by the ammoniacal solution like the copper, and is also reduced to the metallic state; and whatever other metal is not left behind, oxidized by the nitrate acid, is separated as hydrate (lead bismuth), on treating with ammonia. Any arseniate which may have passed into the ammoniacal solution is not decomposed by the copper.

To Separate Silver from Copper.

Sulphuric acid 1 part, nitric acid 1 part, water 1 part. Boil the metal in this mixture until it is all dissolved, adding fresh liquid from time to time as the action ceases. When all dissolved, throw in a little salt dissolved in water, stir vigorously, and allow the precipitated silver to settle, when no more precipitate is formed by the addition of salt water, allow to settle, collect and wash the precipitate on a filter and fuse in a crucible.

Silver from Solution.

Silver may be recovered from the spent plating solutions by adding sulphuric acid slowly until effervescence ceases; allowing the precipitated silver to settle, washing with hot water to remove the acid and fusing in a crucible complete the process. Or, if desired to make silver salts, it need not be fused, as the finely divided silver is acted upon by acids to the very best advantage.

Refining Silver.

After having rolled the silver, cut it into narrow strips, and curl it to prevent its lying flat; the pieces are dropped into a vessel containing two ounces of good nitric acid diluted with one-half ounce pure rain water. When the silver has entirely disappeared, add to the two and a half ounces of solution nearly one quart of pure rain water. Then sink a sheet of clean copper into it; the silver will collect rapidly upon the copper, and you can scrape it off and melt it in a bulk.

JEWELERS' ALLOYS.

Specific Gravities.

The following table shows the specific gravities of numerous metals employed in the arts, together with their melting points, malleability, ductility and tenacity.

Metals.	Specific Gravity.	Melting Points.		Order of Malleability.	Order of Ductility.	Tenacity.
		Fahrenheit.	Centig'de			
Platinum	21.40 to 21.50	Infusible except by the Oxyhydrogen blow-pipe.		6	3	274
Gold	19.25 to 19.50	2016°	1102°	1	1	150½
Mercury	13 56 to 13.59					
Lead	11.40 to 11.45	612	322	7	9	27½
Silver	10 47 to 10.50	1873	1023	2	2	187
Bismuth	9.82 to 9.90	497	258			
Copper	8.89 to 8.96	1994	1090	3	5	302
Nickel	8.50 to 8 60	2700	1482	10	10	
Iron	7.77 to 7.80	2786	1530	9	4	549
Tin	7.25 to 7 30	442	228	5	8	34½
Zinc	6.80 to 7.20	773	412	8	7	109½
Antimony	6.75 to 6.80	A little below red heat.				
Arsenic	5.70 to 5.90	Volatilizes before fusing.				
Aluminum	2.56 to 2.60	1300	705	4	6	300

Gold Alloys.

Take 600 grains of the gold-bearing quartz, finely pulverized, and free from sulphurets; mix with 600 grains litharge, and 7 grains charcoal; melt all in crucible of ample size, and set off to cool. Break the crucible when cold, and the gold will be found in a small button under the refuse matter at the bottom. To ascertain the amount of gold in a metallic substance, select a small sample, weigh it, and melt in a small cupel composed of calcined bone ashes. This absorbs the common metal, leaving the gold and siver exposed to view. The resulting button is melted once more in the proportion of gold 1 part, silver 3 parts, and then rolled into a thin ribbon, and boiled in nitric acid, which dissolves out the silver, and leaves the gold pure at the bottom. The gold can be removed, and the silver subsequently precipitated with salt. In the assay of the rock containing pyrites, it must be roasted until it ceases to evolve sulphurous fumes, then mix 600 grs. of the powder with 300 grs. carbonate of soda, 300 grs. charcoal, 300 grs. litharge, 300 grs. dried borax, and 15 grs. charcoal; melt all in a crucible, and treat as directed above.

To Melt Gold.

Prepare a good fire and heat the ingot in which you wish to cast the gold a little hotter than boiling water; next put the alloy into the crucible, and add a small quantity of pulverized borax, and leave on the fire until melted. Cast this into a clean ingot, and after breaking the bar into small fragments, return to the pot and re-melt the gold, not adding borax this time, but when the gold looks clear and smooth on top, add, for every 6 ounces gold, a piece of saltpeter about the size of a pea, and in about a minute pour the gold. Keep up the heat after adding the saltpeter, and previous to pouring the gold, pour a few drops of gold into the iron ingot. If the stock was clean when you commenced, the gold will roll well. Much depends upon the first rolling of the stock; 18 karat should be subjected to a very heavy strain the first and second draughts, which imparts a grain to the stock; light draughts stretch the gold on the surface, and the middle portion, remaining as cast, causes the gold to crack, many good bars having been condemned, while the trouble was in the rolling. After the 18 karat has been rolled to about twice its length, it must be annealed, then rolled to the size you require. Proceed with melting 14 karat, as above described for 18 karat, giving it as heavy strains in the rolls, but not rolling so much before annealing as the 18 karat. The other karats of cheaper grade do not require the use of saltpeter to toughen; instead of which use a little sal ammoniac, and then proceed as above. When you anneal red gold do not quench it when red hot, but allow the gold to blacken before quenching, otherwise it will slit or seam. Melt new alloys in every case twice; treat solder the same way, to insure a thorough admixture of the copper with the gold.

Coloring Gold Alloys.

Jewelers use a number of different colored alloys for purposes of ornamentation, so as to produce a number of different shades of color in the same article. For example red, blue, yellow and white are employed for flowers; green for leaves; yellow for stems; gray for backgrounds, etc. These are used either made solid and hard soldered, or sweated on the article to be ornamented; it is also used in plating baths and deposited electrically upon the article, the various parts not deposited upon, being covered with wax or asphaltum while plating is being done. For white, silver or platinum are used, the other colors, being made according to the various formulas given below.

Blue Gold.

1. Melt 250 parts of gold and add 250 parts of steel.

2. Melt 500 parts of gold and add 250 parts of iron.

3. Melt 750 parts of gold and add 250 parts of iron.

4. Melt 90 to 99 parts of copper and add 1 to 10 parts of gold. This is the celebrated Japanese blue gold.

Red Gold.

1. Melt 666 parts of gold; add 66 parts silver and 268 parts copper.

2. Melt 750 parts gold, 104 parts silver and 146 parts copper.

3. Gold, 600 parts; silver, 200 parts; copper 200 parts. A very pale shade of red.

4. Gold, 583 parts; silver, 42 parts; copper, 375 parts. Intensely red.

5. Melt 75 parts of gold and 25 parts of copper; incorporate thoroughly by stirring.

6. Equal parts of gold and copper, melted and thoroughly incorporated by stirring form a dark red.

7. Melt 25 parts of gold and 75 parts of copper. This forms a very cheap but effective dark red.

Pure gold varies from a pale yellow to a light copper color, according to the locality in which it is found, and the state in which it occurs in nature, as nuggets, dust, placer gold, quartz rock, etc. In making the alloys care should be taken to use the shade of gold nearest to that desired. Although it is not absolutely essential to do this, much purer colors will be obtained than where this precaution is neglected.

Green Gold.

1. Melt 750 parts gold, 125 parts silver, 125 parts of cadmium. Incorporate thoroughly, remelting if necessary.

2. Melt 750 parts of gold, 166 parts of silver, 84 parts of cadmium.

3. Melt 746 parts of gold, 114 parts of silver, 97 parts of copper, 43 parts of cadmium.

4. Melt 190 parts gold and 50 parts of silver. This has a beautiful green shade.

Gray Gold.

1. Melt together 857 parts of gold, 86 parts of silver, 57 parts of iron or steel.

2. Melt 800 parts of gold and add 200 parts of steel.

3. Melt 725 parts of gold and 275 parts of silver.

4. Melt 833 parts of gold and 167 parts of iron.

Yellow Gold.

1. Melt 583 parts of gold, 125 parts of silver, 292 parts of copper. Dark yellow.

2. Melt 900 parts of gold, 100 parts of copper. Forms a deep yellow.

3. Melt 530 parts of gold, 250 parts of silver, 220 parts of copper. Deep yellow.

4. Melt 666 parts of gold, 194 parts of silver, 139 parts of copper.

5. Melt 750 parts of gold, 146 parts of silver, 104 parts of copper.

6. Melt 666 parts of gold, 333 parts of silver. Pale yellow.

7. Melt 91.67 parts of gold, 8.33 parts of iron. Pale yellow.

8. Melt 91.67 parts of gold, 8.33 parts of silver. Pale yellow.

9. Melt 50 parts of gold, 50 parts of silver. Very pale yellow.

10. Pure gold of the early placer diggings, such as is found in old coins, old jewelry, African placer gold, etc., is generally of a fine yellow, but very soft.

Imitation Gold Alloys.

1. Pure copper, 100 parts; zinc, or preferably, tin, 17 parts; magnesia, 6 parts; sal ammoniac, from 3 to 6 parts; quicklime, 1/8 part; tartar of commerce, 9 parts. The copper is first melted, and the magnesia, sal ammoniac, lime and tartar are then added separately and by degrees, in the form of powder; the whole is now briskly stirred for about one-half hour, so as to mix thoroughly, and then the zinc is added in small grains by throwing it on the surface and stirring till it is entirely fused; the crucible is then covered and fusion maintained for about thirty-five minutes. The surface is then skimmed and the alloy is ready for casting. It has a fine grain, is malleable, and takes a splendid polish. It does not corrode readily, and is an excellent substitute for gold for many purposes. When tarnished, its brilliancy can be restored by a little acidulated water. If tin be employed instead of zinc, the alloy will be more brilliant. It is very much used in France, and must ultimately become popular.

2. Copper, 79.7 parts; zinc, 83.05; nickel, 6.09, with a trace of iron and tin. Called oreide.

3. Copper, 65.50 parts; platinum, 32.02 parts; silver, 2.48 parts. This alloy has about the color of 9 karat gold. Strong boiling in nitric acid has apparently no effect on it even when left in the acid for some time.

4. Fuse with saltpeter, sal ammoniac, and powdered charcoal; 4 parts platinum; 2 1/2 parts copper; 1 part zinc; 2 parts tin; 1 1/2 parts lead. Will stand a close examination without detection.

5. Copper, 11 parts; zinc, 2 parts. Has a rich, deep gold color, is extremely malleable and is employed in making Dutch gold leaf.

6. Copper, 16 parts; platinum, 7 parts; zinc, 1 part, fused together. Has about the color of 16 karat gold and will resist cold nitric acid.

Platinum Alloys.

Several alloys containing platinum, of a comparatively inexpensive nature, have been extensively used under the name of platinum bronze. They are not tarnished by the action of air or water, take a high polish and retain their luster for a long time. The great increase in the price of platinum, owing to its extended use in electrical work, for resistances, etc., will probably diminish its use in the arts for a time, unless new sources of supply shall cheapen it again; but as any jeweler is likely to encounter that which is now in existence, we give several of the formulas for these bronzes:

1. Nickel 900 parts, platinum 9 parts, tin 90 parts. Used for tableware.

2. Nickel 865 parts, platinum 5 parts, tin 130 parts. Used for jewelry and statuettes.

3. Nickel 710 parts, platinum 145 parts, tin 145 parts. Used for tubes of optical instruments.

4. Nickel 316 parts, platinum 32 parts, brass 652 parts.

5. Copper 96 parts, platinum 4. Malleable, rose colored and exhibits a fine grained fracture.

6. Equal parts by weight of copper and platinum. Is ductile, easily worked, has the color and specific gravity of gold, but tarnishes on exposure to the air.

Aluminium Alloys.

Aluminium, or aluminum, is an extremely light, ductile and malleable metal, which is rapidly coming into favor for many purposes since the great improvements in its manufacture and the consequent reduction in cost. It can now be purchased in quantities at ninety cents per pound, which makes it nearly as cheap as copper, when the great difference in weight of a cubic foot of the two metals is considered. It is silvery in appearance, melts at 1,300 degrees F., has a specific gravity of 2.56 to 2.60, which is one-fourth the weight of silver, does not oxidize readily and resists most acids and alkalies, but is very easily attacked by others, especially when heated, or when present during chemical reactions, on other metals. It is three times as ductile as silver, and has 50 per cent. more tenacity or strength. Much nonsense has been written about this metal, such as that it is stronger than steel; will not rust; is not attacked by acids, etc., all of which are untrue. It is readily attacked by many chlorides, such as common salt, (chloride of sodium), etc., and by some of the organic acids, in which respect it resembles silver. In regard to the hardening, tempering, etc., of the pure metal, comparatively little is known at present; but it is probable that as its use becomes more common it will be greatly improved in these respects, as has been done with iron. At all events, it will have an extended trial in the fine arts and mechanics, and it will probably displace platinum and nickel in the various alloys to a large extent, on account of the great difference in weight. One great difficulty remaining to be overcome is that of soldering. At present it can be soldered only by using an alloy of which aluminium forms a part. Several of these solders are given below.

Aluminium forms alloys with many metals; those with copper, silver and tin are largely employed for many purposes, and their use is rapidly extending. The most important are those copper, with which aluminium easily unites.

Aluminium and Gold.

1. Gold 99 parts, aluminium 1 part. A very hard but not ductile alloy possessing the color of green gold.

2. Aluminium 10 parts, gold 90 parts. White, crystaline and brittle.

3. Aluminium 5 parts, gold 95 parts. Brittle as glass.

4. Aluminium 7½ parts, gold 2½ parts, copper 100 parts. Resembles gold in color, and is much used as a substitute for it in jewelry.

5. An alloy of aluminium 999 parts, gold 1 part, is as ductile as pure aluminium, but not as hard as the alloys of silver.

Aluminium and Silver.

1. Aluminium 97 parts, silver 3 parts. Has a beautiful color and is easily worked and is very valuable for articles in which one of the main objects is to obtain lightness, such as the instruments used for marine observations. Those parts of such instruments which, if made with other metal, would weigh four pounds will, when made of the above alloy, only weigh one pound.

2. Equal parts by weight of aluminium and silver give an alloy as hard as bronze.

3. Aluminium 100 parts, silver 5 parts. Can be worked like pure aluminium, but is harder and takes a very high polish.

4. Aluminium 95 parts, silver 5 parts. Is white, elastic and hard. Used for blades of dessert and fruit knives.

Aluminium and Copper.

1. Aluminium 95 parts, copper 5 parts. Patented in the United States by Lange & Sons. It is malleable; can be tempered, and is used for clock springs.

2. Aluminium 10 parts, copper 90 parts. This is used more than any of the other aluminum bronzes. It is hard, ductile, resembles gold, takes a high polish, does not tarnish readily, gives sharp castings, may be rolled in sheets, is more easily worked than steel, and may be engraved. It is used for many purposes in the arts. It can only be soldered with an aluminum alloy.

Aluminium and Tin.

Tin and aluminium give brittle alloys when they contain much aluminium and little tin, but those with a small quantity of aluminium are very ductile and may be used as substitutes for tin.

1. Aluminium 3 parts, tin 100 parts. Hard and not affected by acids.

2. Aluminium 5 parts, tin 95 parts. Forms a very useful alloy for many purposes.

3. Aluminium 50 parts, tin 50 parts. Bourbonne's alloy. This solders easily.

4. Aluminium 100 parts, tin 10 parts.

Aluminium and Zinc.

These alloys are very hard and take a beautiful polish, but owing to their brittle and crystalline nature, when much zinc is present, they are but little used. The most useful is aluminium 97 parts, zinc 3 parts. This is as white as the pure metal, ductile and quite hard. The others are used chiefly as solders.

Aluminium Solders.

1. Melt 20 parts of aluminum in a crucible, add gradually 80 parts of zinc, stirring with an iron rod. When melted, add some fat, raise the heat until fat ignites, then pour in iron molds.

2. Aluminum 15 parts, zinc 85 parts. Melt as above.

3. Aluminium 12 parts, zinc 88 parts. Melt as above.

4. Aluminium 8 parts, zinc 92 parts. Melt as in the first instance.

When soldering dip the iron into a flux composed of copaiba balsam 3 parts, Venice turpentine 1 part, lemon juice a few drops.

Manganese Alloys.

Manganese has a great affinity for oxygen, and hence has the quality of toughening other metals when present during their melting, by destroying the sub-oxides. It is used with many single metals and also in many alloys. Those used for jewelers' tools etc., are:

1. Manganese 25.50 parts, copper 54.50 parts, zinc 20 parts.

2. Manganese 13 parts, copper 55.50 parts, zinc 31.50 parts.

3. Manganese 22.25 parts, copper 52.25 parts, zinc 25.50 parts.

These alloys can all be rolled at a red heat. If it is unnecessary to roll them, iron may be added with advantage, for castings etc., as follows:

4. Iron 5.88 parts, manganese 26.35 parts, copper 56 parts, zinc 11.77.

5. Iron 5 parts, manganese 20 parts, coppor 57 parts, zinc 11.50 parts, nickel 6.50 parts.

Alloys 4 and 5 are nearly white and are much used in cheap plated ware, jeweler's tools, etc.

Silver Assay with Testing Tubes.

Place in the tube enough of the pulverized mineral to fill one inch of the space, and on this pour nitric acid in quantity to occupy 2 inches more, and hold the mixture over a flame until the acid boils. The acid will dissolve whatever silver may be present, and must be passed through filtering paper to remove extraneous matter and return to the tube. Next add a few drops of water saturated with salt; any silver or lead that may be present will be precipitated in a cloudy form to the bottom. Drain off the acid, place the percipitate in the sunlight, and in a few minutes, if it contains silver, it will turn to a purple color, and may be again liquified by the addition of spirits of ammonia. The testing tube is formed of thin glass, about 5 inches long, and less than 1 inch diameter; bottom and sides of equal thickness. Where the tube is lacking a cup may be used instead.

Silver Assay by Smelting.

If no lead is present, mix 600 grs. of the pulverized ore with 300 grs. carbonate of soda, 600 grs. of litharge, and 12 grs. charcoal in a crucible, add a slight coal of borax over all, put on the furnace, melt, take off, give it a few taps to settle the metal, let it cool and remove the button.

Fictitious Silver.

1. Silver 1 oz., nickel, 1 oz. 11 dwts, copper, 2 oz. 9 dwts.

2. Silver 3 oz., nickel, 1 oz. 11 dwts., copper, 2 oz. 9 dwts., spelter 10 dwts.

Imitation Silver Alloys.

1. Copper 64 parts, tin 3 parts.

2. Copper 75 parts, tin 25 parts. Called white tombac.

3. Tin 85½ parts, antimony 14½ parts. Called argentia. Used for spoons and forks.

4. Copper 50; nickel 26; zinc 24. Closely resembles silver and takes a high polish.

5. White copper. See Nickel Alloys 7 and 8.

6. Copper 56; nickel 24; zinc 16; tin 3; iron 2. Used extensively for spoons. Sold as German plate.

7. Copper 100 parts; nickel 70 parts; aluminium 1 part; tungslate of iron 5 parts. Called Minargeul.

8. Tin 87.50 parts; nickel 5.50 parts; antimony 5 parts; bismuth 2 parts. Called Trabak metal.

9. Tin 10 parts; bismuth 7 parts; nickel 7 parts; cobalt 3 parts. Called Warne metal.

See also the various nickel and aluminium alloys.

Nickel Alloys.

1. Copper 40; zinc 32; nickel 8. Called Albata.

2. Copper 60; zinc 30; nickel 10; iron a trace. Called Alfenide.

3. Copper 15; zinc 70: nickel 6. Called White Argentan.

4. Copper 50; nickel 20; zinc 30. German silver; very malleable and takes a high polish.

5. Copper 50; nickel 26; zinc 24. Closely resembles silver. Takes a high polish.

6. Copper 50; nickel 50; zinc 50. Very white and malleable and takes a high polish. Recommended as a substitute for silver.

7. Copper 30; nickel 36; zinc 34. Said to be the Chinese formula for white copper.

8. Copper 41; nickel 32; iron 2½; zinc 24½. Silvery white, sonorous, malleable and ductile. Takes a high polish. Said to equal the Chinese white copper.

Alloy for Compensation Balances.

Berguet used for his compensation balances the following alloy: Silver, 2 parts, by weight; copper 2 parts; zinc 1 part. First melt the silver, and throw in the zinc, reduced to small pieces, stirring the metals and leaving it on the fire for as short a time as possible, to prevent the volatilization of the latter metal; then pour it out and let it get cold. Melt the copper and add the cold alloy, stirring the three together until intimately mixed, Pour out, cut into pieces, and smelt anew, to obtain a perfect incorporation. Be careful, however, to leave the alloy as short a time as possible over the fire, because the zinc dissipates easily. This alloy is hard, elastic, very ductile, and quickly smelts in the furnace. It does not stand much hammering.

Bell Metals.

1. Copper 72, tin 26½, iron 1½ parts. Used for the gongs of French clocks.

2. Copper 78, tin 22 parts. Used for large gongs, bells, etc.

3. Copper 60, tin 24, zinc 9, iron 3. Used for Kara Kanes or Japanese gongs of the first quality, and also for small bells.

4. Copper 60, tin 15, zinc 3, lead 8. For second quality Kara Kanes.

5. Copper 60, tin 18, zinc 6, lead 12, iron 3. For large bells and third quality Kara Kanes.

6. Copper 72, tin 25.56, silver 1.44. For clock bells.

7. Copper 17, tin 80, bismuth 3. For white table bells.

8. Copper 100, tin 20. Melt under charcoal. Product very fine.

9. Copper 1, tin 3. Soft, for musical bells. Bluish red.

10. Copper 1, tin 4. Ash gray. For house bells.

Black Bronze.

Tin 5 parts, copper 83 parts, lead 10 parts, zinc 2 parts. Castings made of this alloy, when heated in a muffle after finishing, quickly assume a dead black appearance, which is not a scale or coating, but is inherent in the metal. It is much used in Japanese bronze objects of art, statuettes, etc.

Violet Bronze.

Copper 75 parts, antimony 25 parts. Has a violet tinge of variable shade according to the shade of the copper and the proportions of antimony.

SPRINGS.

Balances.

A gold balance is preferable to a steel balance. The latter metal has the advantage of being less affected by alteration of temperature, but, on the other hand, gold is denser than steel and is not liable to rust or magnetize.

Balance Spring.

The study of the balance must be of the greatest importance to the watchmaker, because with it chiefly is he able to control the rate of the watch. The great advantage of an overcoil spring is that it distends in action on both sides, and the balance pivots are thereby relieved of the side pressure given with the ordinary flat spring. The Breguet spring, in common with the helical and all other forms in which the outer coil returns towards the center, offers opportunities of obtaining isochronism by slightly varying the character of the curve described by the outer coil, and thereby altering its power of resistance.

Caution About Breguet Springs.

A Breguet spring should never be applied to a watch with an index. It is perhaps the best form of spring for a pocket watch, having all the properties in action of the cylindrical spring, and the great advantage of flatness in form, but any attempts at producing a good timekeeper with this spring and curb pins will end in failure. And any attempt at getting time in positions by pressing the outer coil of the flat spring against the outer or inner pin is mere jobbing, and, even if successful, would require to be repeated every time the balance had to be taken out. For flat springs with regulators it would be highly advisable to pin a spring into the collet, in order to get the stud hole and curb pins to correspond. The end of the over coil of a Breguet spring should run into hole in the stud before being pinned in, and if the stud is screwed into the cock without the balance it will easily be seen if the jewel hole is in the center of the hole in the spring collet, as it should be. This spring should also be pinned at equal turns.

To Demagnetize a Balance and Hairspring.

Remove roller and hair-spring (if not affected). Put in lathe, hold a magnet quite close at first, revolving the balance rapidly; then *gradually* draw the magnet away to a distance of a foot or more. Keep the speed up all the time.

To Prevent Rust.

It is well known that the rusting of bright steel goods is due to the precipitation of the moisture and the air upon it. This may be obviated by keeping the air surrounding the goods dry, and a saucer of powdered quicklime placed in an ordinary show case will usually suffice to prevent the rusting of the cutlery exhibited therein.

To Prevent Rusty Hairsprings.

Brush the outside of the paper parcel, containing the springs, with olive oil—a small quantity only.

Relation of Mainspring to Barrel.

If we wish to have a mainspring theoretically adjusted, there is no better method than simply to allow one-third empty space, one-third for the barrel arbor and the remainder for the spring. When a spring is at rest on the barrel, at either side of the arbor it should occupy one-sixth of the barrel's inside diameter. If we divide a barrel into sixty equal parts, we shall always see that the barrel arbor is just twenty of these parts. It is a great mistake to have a barrel arbor too small, for when such is the case it is almost sure to break the mainspring if the center is at all stubborn, as is very often the case with the cheap class of mainsprings in the market.

Hardening Gold Springs.

To gold detent, thermometer, suspension and balance springs can be imparted a high degree of elasticity. Rolling hardens them, but they are rendered very brittle thereby. They can be made pliable and elastic, not by hardening, as in the case of steel, but by annealing, care being taken not to exceed a certain degree of heat. The spring may be coiled on a block and placed in a tube, with a smooth steel lid; then heat the tube in the flame of a spirit lamp, and as soon as the steel is of a blue temper, remove the flame and allow the whole to cool.

Isochronism of Balance Springs.

The balance spring, of whatever form, to be isochronous must satisfy the following conditions: Its center of gravity must always be on the axis of the balance, and it must expand and contract in the vibrations concentrically with that axis. When these conditions are secured in a properly made spring it will possess the quality of isochronism—that is, its force will increase in proportion to the tension, and it will not exert any lateral pressure on the pivots. M. Phillips, in his memoir, demonstrates these conditions, and proves theoretically that the terminal curves deduced with the view of satisfying the one condition, verify at the same time the other.

WHEELS AND PINIONS.

To Bush a Wheel.

A watch will frequently stop, because a wheel is improperly centered in itself, whereby one side will gear too deep, the other too shallow, into the pinion driven by it. Such a wheel likely is of the proper size, and has good teeth, but the difficulty is its proper centering, when fitted to its pinion. The following will be found an easy way of correction: Take a piece of lead of about the thickness of a silver half dollar, and clip and file it round so that it will fit into one of the larger steps in a step chuck of an American lathe. Screw it fast into the lathe, and while revolving, center and drill a hole of about the size of a winding arbor. Then with a graver, turn out a recess, the size and a trifle more than the thickness of the wheel, so that it will fit in exact, with its teeth touching the outside of the cut. Drive the wheel from its pinion, and broach out the center, so as to take a bush of sufficient length, which should be firmly riveted in and filed smooth on the lower sides. Turn a small groove around the outside of the cut in the lead, crowd in the wheel, with a burnisher set as a gavel. This fixes the wheel perfectly true on the outside. Now center and drill, leaving a little to be turned and with a fine polished graver, to fit the same pinion. Rivet on, and your wheel is all right.

Scape-Wheel Teeth.

Among other differences between English and Swiss watches are the shapes of their scape-wheels; those of the former are pointed, while the latter employ the socalled club-tooth, and experience has demonstrated that the pointed tooth is better than the latter. The reason that club-teeth for lever scape-wheels, prove to be inferior, is because of the adhesion of the nearly parallel surfaces when all is introduced; the increased inertia also being detrimental. It will be found on comparing the two wheels, that the club-tooth will give the largest vibration of the two without oil, and when the oil is fresh, because there is less drop; but this adhesion, when the oil is thick, together with the increased inertia, more than counteracts its advantages.

Putting Teeth Into Wheels.

To put in teeth in watch or clock wheels without dovetailing or soldering them, drill a hole somewhat wider than the tooth, square through the plate, a little below the base of the tooth. Cut from the edge of the wheel, square down to the hole already drilled; then flatten a piece of wire so as to fit snugly into the cut of the saw, and with a light hammer form a head on it like the head of a pin. When thus prepared, press the wire or pin into the empty space of the wheel, the head filling the hole drilled through the plate, and then projecting out so as to form the tooth; then with a sharp pointed graver cut a small groove each side of the pin from the edge of the wheel down to the hole, and with a blow of your hammer spread the face of the pin so as to fill the groove just cut. Repeat the same operation on the other side of the wheel and finish off in the usual way. The tooth will be found perfectly rivited in on every side and as strong as the original one, while in appearance it will be equal to the best dovetailed job.

To Grind Down Plates or Wheels.

The stoning down of plates or wheels with emery or bluestone is rather a tedious job, especially for him who has much of it to do. It can be made easier, however, by using a little soap. The work is more rapidly performed and finer stoning is obtained.

Butting.

The tendency of pinion leaves to butt the wheel tooth when coming into contact is caused either by the bad shape of the teeth or the leaves, or by using a pinion of an improper size, or by the wheel and pinion being placed at an incorrect distance from each other.

To Remedy Worn Pinions.

Turn the leaves or rollers so that the worn places upon them will be toward the arbor or shaft and fasten them in that position. If they are "rolling pinions," and cannot be secured otherwise, it will be better to do it with a little soft solder.

To Tighten a Canon Pinion.

The canon pinion is sometimes too loose upon the center arbor. Grasp the arbor lightly with a pair of cutting nippers, and by a single turn of the nippers around the arbor, cut or raise a small thread thereon.

Pinion Diameter.

The following are excellent rules for determining the correct diameter of a pinion by measuring teeth of the wheel that seizes into it. The term *full*, used below, indicates full measure from outside to outside of the teeth named, and the term *center* the measure from the center of one tooth to the center of the other tooth named, inclusive. For diameter of a pinion of 15 leaves measure, with calipers, a shade less then 6 teeth of the wheel, full. For diameter of a pinion of 14 leaves measure, with calipers, a shade less than 6 teeth of the wheel, center. For diameter of a pinion of 12 leaves measure, with calipers, 5 teeth of the wheel, center. For diameter of a pinion of 10 leaves measure, with calipers, 4 teeth of the wheel, full. For diameter of a pinion of 9 leaves measure, with calipers, a little less than 4 teeth of the wheel, full. For diameter of a pinion of 8 leaves measure, with calipers, a little less than 4 teeth of the wheel, center. For diameter of a pinion of 7 leaves measure, with calipers, a little less than 3 teeth of the wheel, full. For diameter of a pinion 6 leaves, measure with calipers, 3 teeth of the wheel, center. For diameter of a pinion of 5 leaves, measure with calipers, 3 teeth of the wheel, center. As a general rule, pinions that lead, as in the hour wheel, should be somewhat larger than those that drive,and pinions of clocks should generally be somewhat larger proportionally than those of watches.

JEWELS.

New Jewels.

The bad action of a watch may frequently be traced to imperfect jewels. The repairer should carefully examine every jewel in a watch taken down for repairs, and if he finds one with the hole too large, or out of round, that is much wider in one direction than in another, it should be replaced by a good one, in the following manner: If the depth is correct, notice whether the jewel is above or below the surface of the plate; if it is either, then knock it out and cement the plate or bridge on a chuck in the lathe, being careful to get it on true by the hole lately occupied by the jewel. By means of a burnisher raise the burr that holds the jewel in, and if a jewel can be found of the proper size and thickness, and the hole not to large, it can be readily "rubbed in" with the burnisher; if the hole is too small, it can be opened. The chuck on which the article is cemented should have a hole from a quarter to a half an inch deep in its center. If no jewels can be found of the right size and thickness, select one a little too large, enlarge the hole sufficiently to put the jewel in and then proceed to fasten it. If the jewel is broken, of course the same remarks apply to replacing it with a good one. One difficulty which the watchmaker has to contend with, in selecting a jewel from the indifferent lot supplied by some dealers, is to find one, the hole of which is in the center of the jewel. If a jewel is not true, or rather, if the hole in it is not in the center, it must be cemented into a chuck in the lathe, trued up by the hole, then turned off with a diamond cutter, and the chamfer carefully trued up and polished again; while in the lathe it can be turned down to fit the hole in the setting. The shellac is to be removed from the plate with alcohol. In many instances a chuck will have to be turned up to suit the particular job to be done. Care must be taken in opening, or the jewel will break or chip around the hole. The corners must be carefully rounded by a piece of wire larger than the hole, the end of which is conical. It will take but a moment to do this, but if care is not taken too much will be taken off.

Replace a Broken Foot Jewel.

Remove the broken jewel from the collet or setting; place the collet or setting in one of your lathe-chucks, large enough to hold the same; start in motion, and with a fine-pointed burnisher raise the bezel sufficient to receive a new jewel; select a jewel to fit both pivot and setting, replace in chuck, and with a little larger burnisher close down the bezel on pivot, and your job is complete.

To Test the Quality of Watch Jewels.

Place the jewel on a piece of charcoal, and with the blow-pipe and spirit lamp bring it to a bright cherry red. If the stone is perfect and of the proper density, the heat will not affect it; otherwise, the heat will bring out the imperfections, which can easily be detected with a double lens glass. To ascertain if a jewel hole is perfectly polished, place a piece of white paper on your work board and hold the jeweled plate about two inches above the paper and parallel to it, so as to allow the light to pass between the plate and the paper; shade the jewel with a small ring to prevent the light from reflecting from the top of the stone, and with your double lens glass look straight through the jewel hole to the paper. If it is perfectly polished it will appear to have a fine black ring around the inside of the hole. If the jewel is a ruby or a garnet, use black paper instead of white.

Ruby Pin.

If it is necessary to tighten a ruby pin, set it in asphaltum varnish. It will become hard in a few minutes, and be much firmer and better than in gum shellac as generally used.

Removing Jewels.

Jewels can be removed from full plates by putting the plate into a glass tumbler and pouring on nitric acid. The jewels will become loose and drop out after a little time. Wash the jewels well with a little soda or ammonia.

PIVOTS AND STAFFS.

To Measure Length of Staff.

The proper way to measure for the length of staff is, first, to take off both end stones, fit the balance cock properly to the plate (level, etc.), and screw it fast in its place. Then, with the degree gauge, take the measure from the outside of one hole jewel to the outside of the other one, and to this add the amount of end shake the staff is to have, which gives the exact length of the staff between the extreme ends of the pivots. The length should be such that when one pivot rests against its end stone the top pivot shall come level with the outer surface of its hole jewel, and the same when resting on the other pivot. The end shake should be equal to the distance from the outer surface of the hole jewel to the adjacent surface of its end stone when fastened in place. If this distance is neither too great nor small (the jewels must not touch), the end shake will be correct. A safe way for length is to take the outside measure from the surface of the sink in which the bottom end stone fitting rests, to the top surface of the balance cock. Then, having screwed on one of the end stones, shorten up either or both pivots of the finished staff a trifle, to bring the top end of the other pivot level with the surface of its hole jewel as before explained.

Shape of Pivots.

Pivots must be hard, round and well polished; their shoulders are to be flat, not too large, with ends well rounded off so that they do not wear the cap jewel. The jewel holes must be round, smooth and not larger than is requisite for the free motion of the pivot which is surrounded with oil. Their sides must be parallel to those of the pivots, so that they sustain the pressure of the pivot equally at all points of their length. The holes, if of brass or gold, must have been hammered sufficiently hard, so that the pores of the metal are closed to prevent too rapid a wear. It is well if the oil sinks are of a size that will accommodate a sufficient quantity of oil, which, if too little, would soon dry out or become thickened with the worn-off particles of the metal. The under turnings of the pinion leaves are conical, but in such a way that the thicker part be nearest to the pivot, because by this disposition the oil is retained at the pivot by attraction, and does not seek to spread into the pinion leaves, as is often the case, especially with flat watches in which this provision is frequently slighted.

Friction of the Train Pivots.

It is very important to reduce the friction of the wheel pivots to a minimum quantity, and to make it constant so that the motive power be transmitted with the greatest possible uniformity to the pendulum, which is necessary to enable the latter to maintain its arc of oscillation of the same magnitude. The friction of the pivots is due to the pressure of the motive power and the weight of the wheels. The wheel work nearest the motive power must have strong pivots so that they possess sufficient resistance, neither wear the pivot holes to one side nor enlarge them, by which the friction would be increased and at the same time alter the true point of engagement. In tenor with the distance of the wheels from the motive power, the thickness of their pivots must decrease because these latter sustain less pressure, and are subject to a greater velocity than the first parts.

Barrel Arbor.

In the absence of a suitable tap or screw plate, when turning in a Swiss barrel arbor, if the collet is good it may be used as a plate. Soften the collet and file two slight passages across the threads with a fine three-cornered file; screw a piece of brass wire through the collet, so as to free the threads from burr; then re-harden the collet and cut the screw on the arbor with it. A pair of pliers with faces curved to suit the collet are used to hold it. In an emergency the old arbor may be prepared for use as a tap if the old collet is not available.

DRILLS AND DRILLING.

Diamond Drills and Gravers.

Drill a hole or file a notch in the end of a piece of brass wire to correspond with the fragment of diamond; heat the end in a spirit lamp and lay it on a piece of good sealing wax or shellac. When this commences to melt, set the diamond in position and leave the whole to cool. Diamond drills are very commonly mounted at the end of a pin that has had it point filed off; mark a point on the end with a graver and drill the hole, which should be very shallow. Holding the pin in a pin vise, with its point projecting about one-tenth of an inch, heat the vise in a lamp, and proceed as above explained.

To Drill into Hard Steel.

Make your drill oval in form, instead of the usual pointed shape, and temper as hard as it will bear without breaking; then roughen the surface where you desire to drill with a little diluted muriatic acid, and, instead of oil, use turpentine or kerosene, in which a little gum camphor has been dissolved, with your drill. In operating, keep the pressure on your drill firm and steady; and if the bottom of the hole should chance to become burnished, so the drill will not act, as sometimes happens, again roughen with diluted acids as before; then clean out the hole carefully and proceed again.

The Rose Cutter.

The rose cutter is quite a valuable adjunct to a lathe, and is fixed to the spindle in the same manner as a chuck, and will be found exceedingly useful for quickly reducing pieces of wire for screws, etc., to a gauge. For screws, the wire should be of a proper size for the screw heads, and a cutter selected with a hole the size of the finished screw. The point of the wire is rounded to enter the hole of the cutter, against which it is forced by the back center of the lathe, the serrated face of the cutter rapidly cutting away the superfluous metal, the part intended for the screw passes into the hole in the cutter. Some care is required in rounding the point of the wire, for if not done equally all around, the screw will not be true to the head.

To Drill Enamel Dials.

You may have to drill or broach holes in enamel dials. For this purpose use a flat ended drill or concial broach of copper, into which diamond powder has been hammered. A graver kept moistened with turpentine is sometimes used. The edges of the holes in dials may be trimmed with corundum sticks, to be obtained at material shops.

Carbolic Acid on Metal Cutting Tools.

Carbolic acid is recommended for moistening the tools with which metals are worked. The efficiency of the grindstone is even said to be increased by the use of the acid. The dark and impure acid can be used for this purpose.

To Drill Pearls.

The easiest way to hold pearls, in order to drill and otherwise cut them, is to fit them loosely in holes bored in a piece of wood. A few drops of water sprinkled about the holes causes the wood fibers to swell and hold the pearls firmly. When the wood dries they fall out.

MISCELLANEOUS.

Pivot Holes.

A pivot hole is always broached from the inside until the pivot fits; then use a chamferer upon the plate to give the necessary shake. If the underlay in riveting was a nicely polished anvil, and the shake suffices, it is not necessary to chamfer. Then make the oil sink from the outside.

Square Holes.

To file a square hole, it is necessary to reverse the work very often; a square file should first be used, and the holes finished with either a diamond-shaped file or a half-round. This leaves the corners square, as they properly should be.

To Broach a Hole Vertically.

It is quite a serious thing for young watchmakers to broach a hole vertically; a hole in a plate, for instance, that in a barrel, is seldom maintained at right angles to the surface, when they have occasion to employ a broach. They may be certain of success, however, by adopting the following method: Take a cork of a diameter rather less than that of the barrel or other object operated upon, and make a hole in the length of the cork through which the broach can be passed. When the cock has turned quite true on its end and edge, the broach is passed through, and used to enlarge the hole; by pressing against the back of the cock, it is kept against the barrel, whereby the broach is maintained in a vertical position.

To Fit a Bush.

After repairing the pivot, a bush is selected as small as the pivot will admit. Open the hole of the plate or cock so that the bush, which previously should be lightly draw-filed at the end, will stand with a slight pressure upright in the opened hole of the plate or cock; then, with a knife, cut it across at the part where it is to be broken off so that it may break very readily when required to do so. Press it in the plate on the side the pivot works, break off, and then drive it home with a small center punch. In every repair of this nature, notice should be taken of the amount of end shake of the pinion, and allowance made by leaving the bush so that any excess may be corrected. To finish off the shoulder end, a small chamfering tool should be used. It has a hole smaller than the pivot one to receive a fine brass wire, serving as a center to prevent the tool from changing its position while being used; or the wire may be put through the bush holes, and the hole of the tool left open. The above is a far more expeditious way than using the lathe.

The Functions of Oil Sinks.

Oil sinks are formed in watch and clock plates so that by capillary attractions the oil is kept close to the pivot instead of spreading over the plate, and back slopes are formed on the arbors so that the oil may not be drawn all up the body of the arbor. The "attraction" is sometimes negative and becomes a repulsion, as is the case with mercury in a glass tube. It is still called capillarity, whether the fluid is raised above its natural level or depressed below it.

With regard to oil sinks, the views of watchmakers differ. One is in favor of large, the other of small sinks.

Needlessly large, flat oil sinks, as well as projecting bushes, both labor under disadvantages; the former of permitting to escape, and the latter of attracting the oil from the spots where it is wanted. With large sinks the oil spreads very easily and becomes contaminated with dust and fibers, while projecting bushes prevent the oil from again returning to the pivots.

To Remove Broken Screws.

Any one having an American lathe, can, with small expense of time and labor, make a small attachment which will easily and quickly remove a broken screw from the plate or pillar of any watch.

Take two common steel watch keys having hardened and tempered pipes—size, four or five—having care that the squares in each are of the same size and a good depth. Cut off the pipes about half an inch from the end; file up one of these for about half its length, on three equal sides, to fit one of the large split chucks of the lathe. Drill a hole in one of the brass centers of the lathe of sufficient size and depth, into which insert the other key-pipe, and fasten with a little soft solder. Soften a piece of Stub's wire, to work easily in the lathe, and turn down for an eighth of an inch from the end to a size a little smaller than the broken screw in the plate; finish with a conical shoulder, for greater strength, and cross-file the end with a fine slot or knife-edge file, that the tool may not slip on the end of the broken screw; cut off the wire a half inch from the end, and file down to a square that will fit closely into one of the key-pipes. Make a second point like the first one and fit to the other key-pipe, harden in oil, polish, and temper to a dark straw color. Fit the brass center into the tail stock. To use, put the tools in place in the lathe, place the broken end of the screw against the end of the point in the lathe head; slide up the back center and fasten the point firmly against the other end of the screw, that it may not slip or turn; revolve the plate slowly, and the broken screw, being held fast between the two points, will be quickly removed. To remove a broken pillar screw: Place the broken screw against the point in the lathe-head, holding the plate firmly with the right hand, the pillar on a line with the lathe center; turn the lathe-head slowly backward with the left hand, and the screw will be removed. Should the tool slip on the broken screw, and fail to draw it out, drill a hole in the pillar from the lower or dial side, down to the screw point, (if the size of the pillar in the plate will admit of so doing), and with the second point in the back center, remove the screw in the same manner as the plate screw in the first process. Five or six sizes of these points will be found sufficient for a majority of these breakages that may occur.

Cutting Screw Threads.

It is quite a knack to make a nice screw, and beginners are generally apt to use too much force when cutting the thread. If the spindle has been turned too large for the hole in the screw-plate there is danger of breaking the screw-plate, which is over hard, and pieces will chip off; again, the piece to be tapped is apt to break and stop up the hole in the plate, thereby entailing the tedious job of drilling the piece out and cleaning the thread. It is better to begin with a hole much too large and working down gradually. It is natural that a certain amount of force must be employed, and a little practice will soon teach the beginner how much to insure a full good thread. Now, put the screw back in the lathe, and turn the head a little more than the required thickness, and cut the screw off by turning a groove out.

Fitting the Hands.

If the body of the canon pinion will not bear turning in fitting it to the hour wheel, the hour wheel should be opened in the mandrel, as it can not be kept true by opening the hole in the fingers. Fitting the hands to a watch deserves more care and attention than are generally given to it. The way hands are commonly fitted to watches is bad in principle. The pipe of the hour wheel is left too long, and that of the minute hand too short, and when the end shake of the hour hand is adjusted, as it usually is, lay the boss on the hour wheel and the dial, the end shake of the center wheel affects it, sometimes giving it too much and bending the hour hand by its catching the minute hand either in setting the hands or in the going of the watch. In fitting the hands, the examiner should fit the glass, if to a hunting case, as high as the case will admit, ascertain the space available by placing a piece of beeswax on the dial and pressing the glass down on it, and turn the canon pinion until it projects from the dial the height of the beeswax; the hour wheel pipe should rise just perceptibly above the dial, and the end shake of the hour hand be adjusted by the pipe of the minute hand and that of the hour wheel.

Concerning the Rate of a Watch.

When the watch is wound up, and has a lame motion and at last comes to a stop, either the male stop has wedged itself under the female stop in the last winding, or, if it should be a little smaller, then the former applied to a shoulder of the latter, or, the spring is too broad by one number, and, when the barrel cover is pressed in a little, a friction will occur, or finally, the hole is not located in the center of the spring. The first coil of the spring in this case comes to a stand above the spring core, in consequence of which the inner end of the spring rubs on the cover or bottom of the barrel, whereby the spring is pinched in, and thus contracts its elasticity, which has happened to me several times. I took out the spring, filed the two inner sides of the spring a little smaller, rounded their edges with the graver, and with a round file filed the hole in the center. It also happens sometimes that the spring hook is not located exactly in the center of the spring core; there are also springs of an inferior quality, that lose their elasticity after several weeks or months, whereby the watch makes a lame motion and stops after from 10 to 12 hours. It may also occur that the balance spring works loose, which defect also produces a slow motion and finally stoppage; this will occur only, however, when the watch has been wound too quickly, whereby the balance is transported into unduly large vibrations.

To Detect a Magnetized Watch.

Every watchmaker should have a fair-sized pocket compass placed on or gummed to the under side of the showcase glass, and to try every watch that comes in, in the presence of the customer. Place the watch a little to the east or west of the compass, and revolve it slowly; if the watch is magnetized, the compass will move to the right and left as the watch is revolved; if it is not, the compass will remain stationary, while the watch is kept due east or west of the same.

Repairing Cylinder Watches.

It frequently happens that the cylinder edges are worn off, and it does not pay to put in a new cylinder; the watch may, nevertheless, be put into keeping a good rate by altering the escapement. Look at the cylinder and see if there is room either above or below the old wears to shift the action of the wheel. If the wheel holes are of brass, make one a little deeper, and put a shallower one on the other side—this may perhaps be sufficient. This must be done according as you want your wheel up or down. If the holes are stone, shift your wheel on the pinion by a new collet, or turning away more of the old one, as the case may require. If you raise your wheel, see that that it works free of plate and top of cylinder, and that the web of wheel clears the top of passage. This last fault may be altered by polishing the passage a little wider, if the rub is slight. If shifted downward, see to freedom at the bottom of cylinder, etc.

Reducing Diameter of a Watch Glass.

The diameter of a watch glass can be reduce by centering in a lathe, chucking it between two pieces of cork, or a pair of cork arbors, and applying a moistened piece of glass to the edge, or an emery stick. When the desired diameter is attained, polish the edge with pumice stone, followed by putty powder applied on a wet cork.

To Remove Name from Dial.

Take a little diamond powder made into a paste with fine oil, on the end of a copper polisher, the surface of which has been freshly filed and slightly rounded. On rubbing the marks, they will be seen to rapidly disappear. The surface is left a little dull; it may be rendered bright by rubbing with the same powder mixed with a greater quantity of oil, and applied with a stick of pegwood. Operators will do well to previously experiment with several degrees of fineness of the powder on old dials.

INDEX.

Spotted something that isn't quite right?

Let us know and we'll fix it for future reprints. And as a thank you for your diligence, we'll send you a free Sothis book of your choice.

publisher@sothis.press

This page intentionally left blank

Made in United States
Orlando, FL
06 November 2026

90301617R00079